Istanbul, Open City

Urban theory traditionally links modernity to the city, to the historical emergence of certain forms of subjectivity and the rise of important developments in culture, art and architecture. This is often in response to technological, economic and societal transformations in the nineteenth- and early twentieth-centuries in select Euro-American metropolises. In contrast, non-Western cities in the modern period are often considered through the lens of Westernization and development. How do we account for urban modernity in "other" cities?

This book seeks to highlight cultural creativity by examining the diverse and shifting ways Istanbulites have defined themselves while they debate, imagine, build and consume their city. It focuses on a series of exhibitionary sites, from print press/photography, cinema/films, exhibitions of architectural heritage, theme parks and museums, and explores the links between these popular depictions through shared practices of representation. In doing so it argues that understanding how the future is imagined through images and interpretations of the past can broaden current theoretical thinking about Istanbul and other cities.

In line with postcolonial calls for a comparative urbanism that decouples understanding of the modern from its privileged association with Western cities, this book offers a new perspective on the lens of urban modernity. It will appeal to urban geographers and historians, cultural studies scholars, art historians and anthropologists as well as planners, architects and artists.

Ipek Türeli is Canada Research Chair and Assistant Professor of Architecture at McGill University, Canada. She holds a PhD in Architecture from the University of California Berkeley, USA. Her published work focuses on visualizations of the city in photography, film, exhibitions, theme parks and museums. She has been awarded several grants and fellowships for this work, by the Graham Foundation, the Middle East Research Competition, the Andrew Mellon Postdoctoral Fellowship at Brown University and the Aga Khan Fellowship at MIT. She is the co-editor of *Orienting Istanbul* (2010), *Istanbul Nereye?* (2011) and guest editor of *International Journal of Islamic Architecture*'s special issue on "Streets of Protest" (March 2013).

Istanbul, Open City

Exhibiting Anxieties of Urban Modernity

Ipek Türeli

Routledge
Taylor & Francis Group
LONDON AND NEW YORK

First published 2018
by Routledge
2 Park Square, Milton Park, Abingdon, Oxon OX14 4RN

and by Routledge
605 Third Avenue, New York, NY 10017

First issued in paperback 2020

Routledge is an imprint of the Taylor & Francis Group, an informa business

British Library Cataloguing-in-Publication Data
A catalogue record for this book is available from the British Library

Library of Congress Cataloging-in-Publication Data
Names: Türeli, Ipek, author.
Title: Istanbul, open city: exhibiting anxieties of
urban modernity / Ipek Türeli.
Description: First Edition. | New York: Routledge, 2018. |
Includes bibliographical references and index.
Identifiers: LCCN 2017009524| ISBN 9781409422112 (hardback) |
ISBN 9781315590196 (ebook)
Subjects: LCSH: Istanbul (Turkey)–Social conditions. |
Sociology, Urban–Turkey–Istanbul.
Classification: LCC DR726 .T85 2018 | DDC 307.76094961/8–dc23
LC record available at https://lccn.loc.gov/2017009524

ISBN 13: 978-0-367-24411-8 (pbk)
ISBN 13: 978-1-4094-2211-2 (hbk)

Typeset in Times New Roman
by Deanta Global Publishing Services, Chennai, India

For my parents,
Gülhan Karaörs Türeli and Alptan Türeli

Contents

Figures

Acknowledgments

The core of this book is based on my dissertation, completed in 2008 at the University of California Berkeley's Department of Architecture, College of Environmental Design, but it has evolved over the years thanks to the contributions of mentors and colleagues, my students at Berkeley, Brown University, and McGill University, speaking opportunities at various academic meetings and venues and also publication opportunities, all of which blew new life into the individual chapters.

I conducted most of my fieldwork in Istanbul during 2005–7 and in the following summers until 2011. Most of the archival work, which involved physically perusing newspapers and popular journals, was conducted at Istanbul Metropolitan Municipality Taksim Atatürk Library. I also benefited from other local and university libraries, and institutional and personal archives such as that of the History Foundation, Chamber of Turkish Architects, and Istanbul Foundation for Culture. I am extremely grateful to the individuals who have allowed me to interview them, in person, by telephone, or by email, about their memories and experiences and who have shared with me their archival material. Most interviews did not end up in the final manuscript as quotations, but nevertheless informed and enriched the project.

The topics I take up in the book—my interests in visual representations of the city—are heavily influenced by my studies of architecture at professional degree programs, first at Istanbul Technical University (ITU) in the first half of the 1990s and later at the Architectural Association in London. Throughout the 1990s, there were lively discussions in Istanbul on a new breed of (private) museums—museums being a proxy for discussing and challenging given or official history. For example, the History Foundation's acclaimed World City Istanbul exhibition in 1996—curated by faculty from Istanbul Technical University at the Ottoman Imperial Mint Building on the occasion of the Habitat II meeting—was an effort to begin an alternative city museum.

At ITU's Department of Architecture, an elective seminar by visiting professor Nurçay Türkoğlu introduced me to media and popular culture studies. In the spring of 1995, in my final design studio there, I studied under Gülsün Sağlamer, who, together with her teaching team, had developed an exceptionally well-organized brief to design a cinema museum. Our preparatory work involved meeting with some of the legends of Turkish cinema, such as Halit Refiğ, and visiting the state-of-the-art facilities of Mimar Sinan University's Cinema-Television Institute, which was founded by Sami Şekeroğlu, and held the Turkish film archive. As I indulged myself in literature on Turkish cinema, I soon realized that such a museum would also be a city museum since most of the films were shot on location in Istanbul. Some years later during my fieldwork, I enjoyed meeting some of the other prominent figures, e.g. Ertem Göreç, Giovanni Scognamiglio, Şakir Eczacıbaşı, Metin Erksan, whom I had met earlier through cinema literature.

At the Architectural Association (AA), I was fortunate to meet and study under the acclaimed British experimental filmmaker Nina Danino as well as the magnetic architect-educator Pascal Schöning, whose students rarely designed buildings but instead made films as architectural projections or proposals. At this time, founders of muf architecture/art Katherine Clark and Katherine Schonfield taught and spoke frequently at the AA on the relationship of film and architecture. And during my Diploma project, under Carlos Villanueva Brandt and Robert Mull who had inherited Bernard Tschumi's unit via Nigel Coates and NATO, with committed interest in "event-cities," I became convinced that "Architecture is not simply about space and form, but also about event, action, and what happens in space" (*Manhattan Transcripts*, 1976–81, Tschumi). This intellectual and urban milieu informed how I think about architecture and the city, beyond the physical fabric of buildings and individual architects.

At Berkeley, Nezar AlSayyad and Greig Crysler transformed my appreciation of the discipline and provided excellent role models through their intense and popular graduate seminars, and as research and teaching supervisors and mentors. Nezar AlSayyad has been the biggest supporter and challenger of my ideas. I thank him for his generosity during the course of this project and constant reminders since its dissertation stage that I need to translate it into a book. With his deep knowledge of the geographic area I worked in, and also some of its scholarly limitations, he constantly pushed me to make my study speak to broader audiences and fields. It was Greig Crysler who encouraged me to broaden my earlier focus from cinematic representations of the city to explore urban representations in multiple sites. His graduate seminars on globalization, museums and national identity were formative for this project, and constituted the seeds of many ideas that are explored in this study. His thorough feedback and high standards, to the degree that I have been able to meet them, made this a more meaningful and conscientious effort. Outside architecture, Deniz Göktürk's interdisciplinary work on cinema, identity, migration and national identity has been pioneering for film studies and inspirational for me. I especially thank her for the interest she took in my work, for her enthusiasm and generous help at every stage of the project—from giving me the idea to examine Ara Güler's photographic work to co-organizing, together with anthropologist Levent Sosyal, the "Orienting Istanbul" conference in the final year of my doctoral studies and coediting in 2010 the book with the same title that emerged out of that conference.

Several professors and colleagues at Berkeley offered precious insights that informed the development of this study in its early stages: Faculty members Chris Berry, Sibel Bozdoğan, Kathleen James-Chakraborty, Andy Shanken; fellow students Gökçe Kınayoğlu, Susanne Cowan, Lynne Horiuchi, Sara Lopez, Avigail Sachs, and the members of the Townsend Center-affiliated Visual Cultures Working Group coordinated by Anne Nesbet. Outside Berkeley, Levent Soysal, Alev Çınar, Nicole Münnich and Brigitte Le Normand, and Burcak Keskin Kozat organized panels; Jale Erzen and Ipek Akpınar organized conferences where I was able to share earlier versions of the chapters presented here. The "Uykusuz" series at the Istanbul Technical University Faculty of Architecture, the World Affairs Council of Northern California, Istanbul Photography Biennale were kind enough to invite me to share parts of this work with their audiences. The biannual conferences of the International Association for the Study of Traditional Environments (in Dubai, UAE and Oxford, UK) were other venues where I was able to test parts of the work as it was developing.

Following my PhD, I was fortunate to be part of the Cogut Center for Humanities at Brown University as an Andrew Mellon Postdoctoral Fellow where each week once of us shared work-in-progress. I am grateful to my colleagues for their generous and always constructive feedback on the earlier versions of my chapters. In addition to providing direct feedback on my writing, their works stimulated me in numerous ways. I thank my postdoctoral cohort: Ipek A. Çelik, Yukiko Koga, Rina Bliss, Bianca Dahl, Betsey Biggs, Ian Straughn, Rachel Price, Adrián López Denis, Lorenzo Benadusi, Syed Nauman Naqvi, Rivi Handler-Spitz, Silvia Valisa, Shiva Balaghi, Stephen Groening, Katherine Smith, as well as Faculty Fellows Dana Gooley, Marcy Brink-Danan and Herve Vanel; invited faculty members, Engin Akarlı and Rebecca Molholt; and of course, the Director of the Center Michael Steinberg. These colleagues' guidance helped my work speak to multiple disciplinary audiences.

My postdoctoral fellowship was hosted by Brown University's Department of History of Art and Architecture, where faculty mentors Dietrich Neumann, Douglas Nickel, Dian Kriz, Evelyn Lincoln and Sheila Bonde were exceptionally welcoming and helpful. Dietrich Neumann encouraged me to develop seminars directly related to my dissertation. The students who took my courses, "Urban Modernity and the Middle East," "Modern Istanbul" and "Miniature Worlds" were wonderfully receptive, engaged and enthusiastic, and helped me develop as an academic and as a teacher. I feel indebted to them. A semester-long Aga Khan Postdoctoral Fellowship at MIT's Department of Architecture allowed me develop the chapter on the conquest panorama. I thank Nasser Rabbat for his support.

The material in the second chapter on photography became the base of an article in *History of Photography* (2010) as well as a book chapter in *Cool Istanbul* (2014); a précis of the second chapter on cinema was published on the Freer and Sackler's website as an online component to *Ars Orientalis* 42 (2012) and parts were used in *World Film Locations, Istanbul* (2012); the third chapter on exhibitions of architectural heritage appeared with the title "Heritagisation of the "Ottoman/Turkish House" in the 1970s: Istanbul-based Actors, Associations and their Networks," in the special double issue of the *European Journal of Turkish Studies* (2014) on "Heritage Production in Turkey: Actors, Issues, and Scales" edited by Muriel Girard; material from the fourth chapter on the miniature park was published in *Traditional Settlements and Dwellings Review* (2004) and *Orienting Istanbul* (2010). The review process for these publications dramatically improved and refocused the chapters. Feedback at invited talks at MIT, University of California Berkeley, University of California Santa Cruz, New York University, University of Illinois Urbana-Champaign, Binghamton University, Bogazici University, McGill University's School of Planning and the Smithsonian's Freer and Sackler Galleries have helped me further improve my arguments. Ipek A. Çelik and Levent Sosyal have graciously provided comments and edits on the most recent versions of the chapters.

Funds for this project came from various grants at the University of California at Berkeley, Graham Foundation for Advanced Studies in the Fine Arts, Middle East Research Competition, Andrew Mellon Foundation at Brown University and Aga Khan Program at MIT. A generous start-up fund at McGill University helped me enlist the copyediting help of David Moffat. My research assistant Hadi Madwar drew the maps that illustrate locations mentioned in several of the chapters. Thanks are also due to my editors for their patience: Valerie Rose, who signed on this manuscript for Ashgate, and after the acquisition of Ashgate, Pris Corbett and Faye Leerink who made possible its completion under Routledge.

The book is dedicated to my parents Gülhan Karaörs Türeli and Alptan Türeli who encouraged and supported both of their children to pursue academic careers. Experiencing the city through frequent visits to construction sites, to their respective architecture and planning offices and, listening to their conversations about urban planning and issues in Istanbul as a child certainly informed this project even before I had the slightest idea to formulate and pursue it. They taught me to cherish challenges and supported my explorations throughout out my extended studies that included four architecture schools in different cities, countries, and even continents. I cannot thank them enough. My brother Ufuk Türeli always set a high bar in academics from early on, one that I could never match, and later in life, as a professor himself, he generously offered valuable career guidance and advice throughout my doctoral studies and beyond. Last but not least, this project would not have been possible without the emotional and intellectual support of my spouse, Gökçe Kınayoğlu, who was there at every stage. He read numerous drafts, gave so many ideas, and influenced the development of the project in such a profound way that it is hard to disentangle his intellectual contributions from mine; and it is impossible to express in words my deep gratitude after many years, several moves and institutions, and two children. His faith in the project sustained mine.

1 Introduction

The future of the city as a collage of pasts

The past circulates in present day Istanbul—not only because historic monuments and artifacts are scattered through the city but also because the past is continuously improvised upon. One of the most striking advertisements celebrating Istanbul's status as a European Capital of Culture (ECoC) in 2010 inserts the image of Haydarpaşa Train Station in place of Atatürk Cultural Center (Atatürk Kültür Merkezi, AKM) (see Figure 1.1). This ad campaign developed two concepts, one for the domestic audience and another for audiences abroad. Campaign images produced for exhibition outside of Turkey were based on the concept of "situating Istanbul on par with other leading world metropoles" and featured a seductive yet brooding silhouette of the historic city produced by photographer Rainer Stratmann. In contrast, domestic campaign images uprooted well-known monuments and placed them in other, also familiar locations. The concept was, purportedly, to rediscover the city. The spokesperson of the Istanbul 2010 ECOC Agency at the time explained: "We wanted [the residents of Istanbul] to become aware of the city. For this reason, we came up with this concept of re-discovery ... We want Istanbulites to recognize the wonders they take for granted."[1] How does such an image, a strange image, actually work? What does it communicate? What are the different ways it could be read?

In order to understand this composite image, a brief story of the AKM, a building around which considerable power struggles have been and continue to be staged, is in order. The AKM was first proposed in the mid-1930s as a performance venue for opera for Taksim Square. After several designs and multiple halts in construction, architect Hayati Tabanlıoğlu designed the final version. Following a fire in 1970, the building was closed and opened again in 1977. In addition to the art galleries located on its top floor, the AKM has halls to stage opera, ballet, theatre, classical music performances, and concerts. It has a commanding position on Taksim Square. Its abstracted façade composition, alluding to a rising curtain, austere massing, elegant modernist detailing, and, most importantly, functional programming, have all aided the building's association with early Republican-era cultural modernism, though it was built much later. The AKM was closed in 2008, allegedly for maintenance and repairs, but its prolonged closure and rumors of planned demolition have triggered much consternation within the architectural

Figure 1.1 Istanbul 2010 Advertising Campaign image shows Haydarpaşa Train Station in place of Ataturk Cultural Center on Taksim Square.

Produced by advertising agency UltraRPM. İstanbul İl Özel İdaresi.

community. Since the mid-1990s, there have been verbal proposals to build a grand mosque on Taksim Square.[2] Coupled with the current government's distaste for and disapproval of Western forms of performance art, the fear of some (and the stated goal of the current President of Turkey, Recep Tayyip Erdoğan) is that the Cultural Center will be torn down and a mosque will be built, changing forever the association of the square with cultural modernism.[3]

Haydarpaşa Train Station, which was montaged into the location of the AKM in the original photograph, is another controversial site (see Figure 1.2). This train station is the terminus of railways from Anatolia. It was built between 1906 and 1908 by German architects Otto Rittner and Helmuth Cuno as a link in the Berlin–Baghdad railway. It represented an important stage in the incorporation of the Ottoman Empire into the economic system of Western powers. It then gained symbolic importance as the setting of arrival and departure scenes to Istanbul in literature and film. Yet, delinked from its physical context and history and juxtaposed with Taksim Square, it simply, and ironically, serves to "Ottomanize" (and by extension Islamicize) this modernist, secular Republican space. Another aspect about Haydarpaşa that ought to be mentioned is that the current government has been plotting to decommission and privatize the train station and the port area behind it for over a decade, despite protests and various forms of activism to protect it as a train station. Thus, when the 2010 ECOC Agency representative suggested that the ad image seeks to enable rediscovery, his words do not reflect the high level of anxiety the image may (and is perhaps intended to) provoke for individuals familiar with the politics of these places.

Figure 1.2 Original photograph used for the photomontage in Figure 1.1.
Photographer Azmi Dölen. İl Özel İdaresi.

Fast-forwarding three years, in 2013, the seemingly successful government of the AKP (Justice and Development Party, Turkish: Adalet ve Kalkınma Partisi), in its third term, and political analysts were taken by surprise by the civil resistance focused on Gezi Park, abutting Taksim Square. The protests were triggered by the violent crackdown on peaceful demonstrators who sought to protect the park from demolition; the government's architectural proposal, which would be allegedly implemented by a company owned by Erdoğan's son-in-law, became the symbol of an increasingly authoritarian and corrupt government. The controversial project aims to privatize the public park to reconstruct in its place a mall resembling an Ottoman barrack that preceded the park—yet another attempt to rid Taksim Square and its surrounds of their representational symbolism as a Republican space of cultural modernization. The irony is that the former barracks were also a cultural modernization project of the late Ottoman state, and had replaced an Armenian cemetery.

Visual representations have power to shape the built environment. They can be read and used to support or critique particular projects in ways that may not have been intended by their original producers. The fictional images of the reconstructed Ottoman barracks used in news reports are a case in point. The images were lifted from an exhibition entitled "İstanbul'da Tarih ve Yıkım: Hayal-et Yapılar" (History and Demolition in Istanbul: Ghost Buildings, 2010), sponsored by the aforementioned ECoC Agency—this was an exhibition which had no intention of promoting the reconstruction of the barracks or, for that matter, any of the demolished buildings exhibited in the show.[4] According to the preparatory archival exploration of the organizers, there had been earlier proposals in the 2000s to rebuild the barracks, but those proposals were shelved due to lack of support in the national assembly. The idea was revisited in 2011 and announced by Erdoğan as one of his election projects with an image from the exhibition.[5] According to one of the architects who worked on the exhibition: "Since no one had any idea about the Taksim project, [Erdoğan and later news reports] used [their] image since it was already imagined."[6] The exhibition organizers were disappointed that their images were used without permission to visualize and therefore advocate Erdoğan and the AKP's agenda; however, it was those same images that enabled different, agonistic, even antagonistic viewpoints to be articulated and framed the debate and informed the outcome of the built environment.

As a result of the protests, the reconstruction of the barracks was halted and the AKM gained new prominence in its abandoned state as the different groups participating in the protests used its plain, dark façade to hang posters and banners. The façade became a "speakers' corner," as it were. One directly addressed Erdoğan and defiantly said: "Shut up, Tayyip." The violent police crackdown on the protests also involved cleaning that façade of the peoples' messages and replacing them with a neatly arranged display of a giant portrait of the Republic's founding leader, Mustafa Kemal, flanked by even bigger flags, temporarily restoring the building's republican symbolism.

The protest for Gezi led to unprecedented unification, not only in response to police brutality but also because the controversy in question was the survival of a public park that could be physically occupied (when all other channels of communication failed), and, second, because there was a "tangible" architectural proposal. The AKP's proposal for Gezi Park and the area around it, including the AKM and Taksim Square, became the focus of a lively debate, not only in Turkey but internationally. The previous decade was marked by a booming arts and cultural scene, the diversity and richness of cultural actors, newfound prosperity made visible in new consumption patterns under a neoliberal regime, and the open-mindedness of the government in supporting all this effervescence. As one analyst argued, the Gezi protests marked the "end of Cool Istanbul" (coined as such on the cover of *Newsweek* on August 28, 2005). Since then, Erdoğan's belligerent response to public dissidence has only worsened.

The proposed replica of a long-gone Ottoman-era barrack, is a demonstration of Svetlana Boym's suggestion, observed in relationship to Eastern European cities, that the future of the city is increasingly imagined based on improvisations on its past.[7] The publicity image, with Haydarpaşa Train Station collaged into Taksim Square to replace the AKM, is a demonstration *par excellence* of the future of the city as a collage of pasts.

Open city

The term *open city* is an expression I came across in popular, mass-circulating dailies of the mid-1950s.[8] It is a useful conceptual tool that allows me to tie seemingly separate phenomena—economic reforms, demographic growth, and changing urban form—to anxieties related to the city's accelerating transformation. Originally used in conventions of war in Europe from the seventeenth century onward, the term designated cities that chose to abandon all defensive efforts, allowing invaders to simply march in, in order to protect historic landmarks and civilians.[9] In contemporary urban planning literature, the term is sometimes invoked to describe a lack of growth controls that might otherwise be used to divert unwanted migrants. As such, "open city" represents an ideal on the opposite side of what Richard Sennett calls the "brittle city"—a closed, over-determined system that denies chance encounters, narrative possibilities, and growth over time.[10] Recently, the term open city has acquired new cachet, prompting architects to consider whether design can be utilized to achieve this political ideal—to explore how "designing coexistence" might be possible.[11] This new interest has been coupled with a return to Henri Lefebvre's conceptualization of the right to the city and the broader movement of urban activism.

Theoretical reflections on the open city have so far focused on how the city can manage cultural complexity and the role manipulations of space and architecture might play in furthering this ideal. Defining the open city as one where "immigration is accepted by the people and the urban politics," Detlev Ipsen remarks: "the conditions required for an open city lead into a contradictory field of ideas, perceptions, and evaluations … What some appreciate as colorful variety, others view as dangerous chaos."[12] This is the paradox of the open city—that it generates simultaneous tolerance and ignorance, ease and anxiety.

Rather than provide causality for a chain of events purporting to explain the city's form, I seek to peel back the layers of cultural anxiety that shape the way the city is experienced today. What are these anxieties? Mainly, Istanbul changed dramatically since the 1950s. It population, which hovered around the one million mark during the first half of the twentieth century, increased from 1,166 million in 1950 to 10,019 million in 2000.[13] Triggered by national population growth, economic reform, and rural-to-urban migration and enabled by urban renewal, the city expanded toward its peripheries through the construction of new speculative and state-subsidized housing developments, frequently surrounded by shantytowns. Physical growth was paralleled by the opening of the city's physical social landscape—an opening that was not necessarily welcomed by all and was simultaneously celebrated, criticized, and mocked. While migration was not a novel phenomenon in the city's long history, when coupled with urban renewal it led to feelings of uncertainty. The image of "invasion" has become a trope of hegemonic public discourse on urban transformations in Istanbul. Underlying such anxiety is the ambiguity of the class identity of newcomers, some of whom were able to quickly acquire access to middle-class lifestyles.

Urban modernity

The concept of *urban modernity* refers to the experience of modern city life and the associated cultural celebration of originality.[14] There are multiple general points of reference for modernity: the post-sacred world of the Renaissance, the intellectual framework of the Enlightenment, and the psychological and social changes of the nineteenth and twentieth centuries connected to industrial modernization. This last conceptualization, however, is essentially an urban phenomenon.[15] Thus, urban theory traditionally links modernity to the historical emergence of certain forms of subjectivity and the rise of important developments in arts and architecture. In particular, these are seen to have arisen in response to technological, economic, and societal transformations in nineteenth- and early-twentieth-century Euro-American metropolises such as Paris, Berlin, and Chicago. According this view, urban modernity is typically also associated with the creation of a more open physical environment through the construction of boulevards and infrastructure to support the movement of people and goods. By showcasing an enlarging consumer economy and enabling class encounter in public spaces, such interventions are thought to advance democratization. In contrast, non-Western cities in the modern period are generally considered through the lens of development.

The study of urban conditions is intimately tied to the geopolitics of privilege; thus, analyses of Istanbul are connected to evaluations of Turkey's international standing. Through much of the twentieth century, Turkey was heralded as an example of a country that had been successful at modernization. But from the 1990s onwards, the country's modernity has come to be seen as flawed. In the immediate post–World War II period, Turkey aligned itself with the new international economic order dominated by the United States and its monetary and military institutions. Market reforms in the 1950s, the import-substitution model of the 1960s and 1970s, and economic liberalization in the 1980s all contributed to different stages of a process by which Turkey followed the lead of the United States while hoping to join an integrated new Europe. Starting with the election of the Democrat Party (1950) in the country's first multiparty or free elections, the Turkish state loosened its grip on the production of a homogenous national culture. At the same time, rapid urbanization brought a new political and social plurality to urban life, and substantial parts of the population were able to transform their economic orientation from subsistence to consumption by moving to the city.

Increased democratization, however, led to alternative projects of modernization that steered away from the old language of cultural modernism. One path has been neo-Ottomanism; sometimes this is driven by nostalgia for the cosmopolitan past of the city, sometimes by yearnings for Muslim-Ottoman supremacy. This has resulted in what Esra Özyürek calls "nostalgia for the modern."[16] The term "modern," as used here, refers to the former Turkish Republican milieu, such that at the turn of the millennium, Turkey seemed split between an uncritical embrace of the early Republican modernist vision and the denunciation of that vision as the byproduct of a repressive regime. Another development has been the repudiation of what Meltem Ahıska calls the "historical fantasy of the modern," conceived as a "time-lag" with the West.[17] This parallels the debate on "alternative modernities."[18] In limbo at the gates of an integrated Europe, the overtly political task of culturally appropriating Istanbul has involved rewriting the city. This book responds to and tries to contextualize the resulting hegemonic narrative of loss that has permeated local literature and urban discourse.

The second half of the twentieth century in Istanbul was defined by rapid urbanization. The city witnessed the rise of a consumer society, the expansion and cultural ascendancy of the middle class, and the development of modern techniques of image production and consumption. At the same time, the "opening" of the city through physical expansion and population influx gave way to widespread concern about the city in the realm of culture which persists to this day. *Istanbul, Open City* is thus an account of urban modernity shot through with nostalgia and marked by an obsession with images and definitions of urban public culture. It addresses a turn-of-the-millennia city burdened with a pervasive feeling of loss in response partially to unplanned rapid growth and Turkey's inchoate position vis-à-vis Europe.

Istanbul, Open City aims to reclaim the lens of urban modernity, somewhat exhausted by the study of a few wealthy Western metropolises, and look with fresh eyes at the city. Istanbul has hitherto been studied using frameworks of Westernization, modernization, Third World urbanization, and, most recently, neoliberal urbanism. In examining Istanbul through the lens of urban modernity rather than modernization or its current iterations, economic globalization and neoliberal urbanism, *Istanbul, Open City* responds to Jennifer Robinson's call to

> [decouple] understandings of the modern from its association with the West, and … dislocate accounts of "urban modernity" from those few big cities where astute observers elaborated on the broader concept of "modernity," placing it in a privileged relationship to certain forms of life in these places.[19]

Thus, the book seeks to highlight cultural creativity by examining the diverse and shifting ways Istanbul residents have defined themselves while debating, imagining, building, and consuming their city.

Economic globalization and neoliberal urbanism have recently become hegemonic in scholarly interpretations of the contemporary city. Meanwhile, scholarship on "urban imaginaries" seems to have evolved separately, privileging the way city dwellers imagine their own environments. Works by David Harvey and James Donald exemplify the opposite ends of this spectrum. In the former, the spectacle of the city is a result of capital over-accumulation; in the latter, the city is an "imagined environment."[20] This book seeks to bridge these two perspectives.

Istanbul as a focus of urban history and urban studies

How do we envision urban history? As an academic field of inquiry, urban history is produced by scholars from a variety of disciplines, including, but not limited to, social history, cultural history, architectural history, planning history, urban sociology, urban geography, and archeology. An architectural historian's urban history is typically the history of the physical fabric, the built environment, and the contributions of architects, and is fundamentally different, for instance, from a social historian's concerned with social change.[21] This book follows a visual culture approach to urban history. It is through visual images that the modern city is typically experienced as an entity. While there can never be a singular vision of urban history in and of any city, predominant or popular perspectives on the built environment are shaped by visualizations of the city. Furthermore, cities can be conceived as imagined communities of individuals who may not have face-to-face contact but maintain a sense of belonging that is fostered by various media.[22] In turn, it is to urban visualizations that scholars can turn in order to understand how a city is experienced collectively and cumulatively. Images of the city produce their own audiences, some more temporal than others. This book is about how the past of the city is summoned to imagine the future, both to the aid of the present. It seeks to explore Istanbul's selected past(s) through its visual images in circulation. It deals episodically with the second half of the twentieth century.

Urban studies dealing with modern and contemporary Turkey from either a historical or a sociological approach draw respectively on theories of nationalism and globalization, and focus either on early nation-state formation (1923–38) or post-1980 developments.[23] Historical urban studies largely concentrate on the early Republican period, with a special focus on the use of education, art, and architecture as tools of nation building. In revisionist history writing, nationalism and modernity are regarded as top-down projects that did not originate from indigenous societal transformations. Sociological urban studies, on the other hand, examine contemporary negotiations of public space. From this perspective, the wearing away of the early Republican project of nationalist modernity is attributed to the forces of neoliberal globalization. Underlying this period-theory coupling is the populist assumption that Istanbul was marginalized until the 1980s, when its fortunes turned around.

The history of Istanbul in the second half of the twentieth century remains an unpopular, perhaps precarious, topic among architectural historians of the city. As the Turkish architectural critic Uğur Tanyeli bluntly argues, "Architectural historiography in Turkey omits/forgets Istanbul."[24] This has to do with the difficulty of dealing with the massive growth of the city that has taken place outside the control of expert visions. Zeynep Çelik's *Remaking Istanbul* (1984) and Murat Gül's *The Emergence of Modern Istanbul* (2009)—both planning and architectural histories—examine the nineteenth and first half of the twentieth centuries, focusing on administrative reforms and certain planned physical changes. An important recent revision is Sibel Bozdoğan and Esra Akcan's coauthored survey *Turkey: Modern Architectures in History* (2012), which summarizes urbanization patterns and actors with due emphasis on Istanbul while maintaining its focus on the works of practicing architects. There are more Turkish-language sources on the architectural and planning history of the city, notably the works of planning historian İlhan Tekeli.[25]

Tekeli suggests four periods in the urban development of Istanbul and other large cities in Turkey: "shy" modernity from the mid-nineteenth century to the proclamation of the Republic in 1923; "radical" modernity to (the switch to multi-party politics in) 1950; "populist" modernity to (the coup of) 1980, which marks the opening and integration of the Turkish economy with global markets; and, finally, the "erosion of modernity" since the 1980s.[26] The postwar era was marked in the West by managed capitalism and planned urbanization. However, the legitimacy of modernist planning was always undermined in Istanbul, and in similar metropolitan regions of developing countries, by the realities of urban-to-rural, migration-driven population growth and housing shortages. The opening of the economy in the post-1980 period, in line with the demands of post-Fordism and the reorganization of the world economy, meant a waning in state entrepreneurship, the end of import substitution industrialization, and integration with global markets. In the urban development of Istanbul, this translated to a change of scale in housing development from individuals or small-scale developers to large built-up areas and new building-supply methods, such as those enabled by the state enterprise TOKİ (Mass Housing Administration) and the reorganization of industrial production in special zones and sites for warehouses, wholesale trade centers, transport services, and specialized production away from the traditional city center. This led, among other transformations, to the move of the central business district to the Levent-Maslak axis, now marked by glass-clad towers, the repurposing of abandoned industrial properties as cultural institutions, and the gentrification of inner-city historic areas for touristic-cultural consumption purposes. Thus, "erosion" may not be an appropriate qualifier for post-1980 developments since what we are observing is but a reassertion of modernization with a vengeance and an intensification of the contradictions of modernity.[27] This book has benefited from the insights of the above-mentioned architectural and planning histories; however, it is not about architects, planners, and their professional activities, nor does it try to provide a comprehensive account of the development of the city's macro form. It eschews in particular a political periodization in favor of media-specific readings.

The city has been a laboratory for sociologists studying Turkey, who have tended to generalize from fieldwork in Istanbul to the rest of Turkey. Early on, their focus was primarily on migration, squatter settlements, and the question of "integration."[28] It was only in the 1990s that studies of the city started to examine changing "lifestyles" brought about by neoliberal economic policies and the efforts of successive governments after the 1980 coup to turn Istanbul into a world city. Çağlar Keyder's edited volume *Istanbul: Between the Global and the Local*, published in 1999, was the first book to bring this new generation of studies together with a cultural perspective and a focus on practices of the middle class. The question Keyder asked in a 1992 essay in the History Foundation's magazine *Istanbul* was whether Istanbul would be able to achieve its potential as a "global city" in the sense of theorists such as Saskia Sassen, or whether it would miss the opportunity that was unfolding in the aftermath of the dissolution of the Soviet Union. In the post–Cold War world, Turkey gained new geopolitical significance vis-à-vis the Middle East and the Central Asian Turkic states. However, it was not clear in the 1990s which path the Islamists then in power—with Erdoğan from the Refah (Welfare) Party as Istanbul's mayor—would take regarding the global city project. Contributions to Keyder's volume reflected this uncertainty about Istanbul's status and trajectory. Yet since 2002 under a single party, the AKP, Istanbul witnessed relative stability and a more formal assertion of neoliberal values.[29]

Contemporary sociological scholarship on Istanbul has been conversant with concurrent debates in Urban Studies—somewhat dominated by the Anglo-American academy. They have taken up issues of fragmentation;[30] gated communities;[31] gentrification in the historic quarters of the city;[32] generalized gentrification of slums;[33] and the revitalization of deindustrialized areas.[34] My coedited book *Orienting Istanbul* (2010, with Deniz Göktürk and Levent Sosyal) explores how processes of creative production and exhibition are intertwined with neoliberal urban restructuring.[35] A number of new edited works have appeared since, unpacking some of the topics *Orienting Istanbul* introduced, ranging from the impact of top-down urban renewal to creativity in the cultural sector.[36]

Unlike these preceding studies, however, *Istanbul, Open City* is primarily concerned with how versions of the past circulate in the present to imagine the future. *Istanbul, Open City*'s contribution to local urban studies aligns more with historical research that calls attention to the importance of public spaces in enabling the creation of the public sphere.[37] For example, in the new breed of (Ottoman) urban history, physical spaces and their visual depictions are no longer treated as expressions of culture or representations of top-to-bottom power, but as agents for political and social negotiation among various public factions and the state.[38] As in these works, this book also uses visual sources as primary material, but to look at the present.

Organization of the book

In this book, I look at a series of exhibitionary sites—photographs for print journalism (Chapter 2), cinema films for popular consumption (Chapter 3), public exhibitions of architectural heritage (Chapter 4), and museums and theme parks (Chapters 5 and 6)—as well as links between them that emerge from shared practices of representation. My use of the term "exhibitionary sites" here alludes to Tony Bennett's "exhibitionary complex," by which he explains how populist sites contributed to state-sanctioned narratives of national culture.[39] Echoing this theme, I examine how Istanbul has been staged according to competing power-effects. Underlying the creation of these sites (each of which transcends physical place to include its mediated transmission) is a desire to shape the public through culture and create urbanite citizens out of crowds. Contemporary debates on the future of Istanbul often also include retrospective analyses and reinterpretations of these same sites.

The chapters can be read independently. The analysis in each chapter calls for a different body of scholarship and the nature of the material has demanded separate framings. Yet, when read together, the reader will see clearly the overlaps and certain continuities between the chapters.

In Chapter 2, I discuss how journalistic street photography documented the newcomers, in whose mirror image it was possible for middle class consumers of newspapers and magazines to define themselves as authentic Istanbulites. In Chapter 3, I examine cinema films, which catered to an even wider spectrum of audiences, to portray the rapid process of urban expansion by housing. I also examine the process of "apartmentization" (preference for multistory modern apartment buildings over traditional old wooden houses) and the struggle of newcomers for a right to a middle-class identity via access to modern housing. The films' urbanite directors tended to warn their audiences about the temptations of the city with didactic messages. By the 1970s, modern multistory apartments became widely accessible, and thus no longer a sign of distinction for middle-class status. Chapter 4 takes up how apartmentization and the newcomers became the culprits behind the disappearance of old wooden houses and the traditional neighborhoods they constituted and, by implication, of "urbanity." Thus, the growth and opening of the city further led to anxieties about the loss of an urbanite public culture and vernacular architectural heritage that might present Turkey's distinctness and worthiness on a European stage.

Starting with exhibitions of the disappearing vernacular fabric, architectural reconstructions and, lately, theme-park environments have emerged as important venues for the manufacturing and consumption of a cosmopolitan and "Old Istanbul," a construct that emerged in reaction to rapid change and urban renewal. Along with literary accounts and pictorial depictions of the past, such sites seek to alleviate anxieties about the city's continued transformation. Chapters 5 and 6 focus on efforts by the AKP-controlled city administration to articulate local distinction on a global stage through the simulacra of select moments and buildings from the city's Ottoman past. These sites also inspire and draw from a broader visual cultural production that includes, among other media, films and TV productions. They seek to evoke the former influence of the city as the seat of an empire, hinged to the idea of conquest by nomadic "Muslim Turks," and as the promise of what it can become in the future.

Using Istanbul as a case, I make several interwoven arguments. I discuss how, with increased democratization and the multiplication of public spheres, urban representations have moved from mediums of technological reproducibility such as photography and cinema to intentionally distorted replicas such as those found in contemporary theme parks. My study thus pursues an analysis based on different "phases of the image," as theorized by Jean Baudrillard in the early 1980s, to explain the movement from representation to simulation characteristic of advanced capitalism.[40] I am not suggesting one medium supplements another one in a successive order but rather that they coexist, and I identify overlaps in techniques of representation. I illustrate how, in the process of staging and consuming a city, both the producers of and audiences for urban representations assume new positions. Chapter 4 on the heritagization of the Turkish house deals most explicitly with this, as it traces how experts and enthusiasts turned into preservationists. Finally, I argue that urban representations are constitutive of reality rather than merely reflective of it; thus, they have particular effects on the experience of a city and the development of its built form.

While chapters follow an episodic and loose chronological order, each reflects on the others and has a bearing on the present—a present heavily burdened with a feeling of nostalgia. I revisit the idea that the future of the city is imagined based on improvisations of its pasts in my analysis of these different exhibitionary sites. All of the exhibitionary sites discussed are co-opted by the city administration or the central government at times. But it is not only the city administration, political associations, or parties who are imagining the future of the city. Whether by organizing exhibitions, circulating images on or from its past, Istanbul enthusiasts, artists, curators, civil society organizations; in consuming these images, ordinary residents of the city also contribute to imagining the city's future.

Notes

1 Chairman of the Executive Board Şekib Avdagiç, quoted in, "İstanbul 2010'a ilginç reklam kampanyası," *Hürriyet*, December 4, 2009, accessed February 28, 2010, http://www.hurriyet.com.tr/gundem/13103553_p.asp.
2 Tim Arango, "Mosque Dream Seen at Heart of Turkey Protests," *New York Times*, June 23, 2013, accessed June 23, 2013, http://www.nytimes.com/2013/06/24/world/europe/mosque-dream-seen-at-heart-of-turkey-protests.html?pagewanted=all.
3 "AKM'yi yıkacağız, Taksim'e cami de yapacağız: Başbakan Erdoğan, Taksim'i yenileme çalışmalarında kararlı olduklarını yineledi," *Dünya*, June 2, 2013, accessed June 23, 2013, http://www.dunya.com/akmyi-yikacagiz-taksime-cami-de-yapacagiz-193887h.htm.
4 Accessed on January 10, 2017. http://www.hayal-et.org.
5 Erkan Aktuğ, "Hayal-et'ti, gerçek olacak!" *Hürriyet*, June 3, 2011, accessed on January 10, 2017, http://www.radikal.com.tr/turkiye/hayal-etti-gercek-olacak-1051623/.
6 Cem Kozar, "Hayal-Et Yapılar Sergisi'nde Taksim Kışlası," *Arkitera.com*, February 29, 2012, accessed January 10, 2017, http://www.arkitera.com/gorus/268/hayal-et-yapilar-sergisi-nde-taksim-kislasi.
7 Svetlana Boym, *The Future of Nostalgia* (New York: Basic, 2001), 75.
8 For example, an editorial by Nihad Sami Banarlı titled "Opening Istanbul," *Hürriyet* on October 6, 1956, celebrates Istanbul's new boulevards, squares, and other physical openings. A caricature on the front page of *Milliyet*, with the title of "Istanbul Open City," on October 20, 1956, mocks the promiscuity and cultural degeneration in the city.
9 Gregory J. Ashworth, *War and the City* (London; New York: Routledge, 1991), 157–8.
10 Richard Sennett, "The Open City," in *The Endless City*, eds. Ricky Burdett and Dejan Sudjic (London: Phaidon, 2008), 290–7.
11 Tim Rieniets, Jennifer Sigler, and Kees Christiaanse, eds., *Open City: Designing Coexistence* (Amsterdam: SUN Architecture, 2009). An exhibition with the title "Open City Istanbul" was curated in 2010 by Philipp Misselwitz and Can Altay and sponsored by the Istanbul 2010 ECoC Agency, in collaboration with the 4th International Architecture Biennale Rotterdam IABR "Open City – Designing Coexistence." However, I have not seen this exhibition. Information available at: http://www.depoistanbul.net/en/activites_detail.asp?ac=25/.
12 Detlev Ipsen, "The Socio-Spatial Conditions of the Open-City: A Theoretical Sketch," *International Journal of Urban and Regional Research* 29, no. 3 (2005): 644–53.
13 Ferhunde Özbay, "İstanbul'da 1950 sonrası Nüfus Dinamikleri," *Eski İstanbullular Yeni İstanbullular*, ed. Murat Güvenç (Istanbul: Osmanlı Bankası Arşiv ve Araştırma Merkezi, 2009), 54–77.
14 Jennifer Robinson, *Ordinary Cities: Between Modernity and Development* (London: Routledge, 2006).

15 David Harvey, "Modernity and Modernism," in *The Condition of Postmodernity: An Enquiry into the Origins of Cultural Change* (Cambridge, MA: Blackwell, 1990), 10–38; especially, 25–8.

16 Esra Özyürek, *Nostalgia for the Modern: State Secularism and Everyday Politics in Turkey* (Durham, NC: Duke University Press, 2006).

17 Meltem Ahıska, "Occidentalism: The Historical Fantasy of the Modern," *The South Atlantic Quarterly* 102, no. 2/3 (2003): 351–79.

18 Dilip Parameshwar Gaonkar, "On Alternative Modernities," in *Alternative Modernities*, ed. Dilip Parameshwar Gaonkar (Durham, NC: Duke University Press, 2001), 1–24; and Andreas Huyssen, "Introduction," in *Other Cities, Other Worlds: Urban Imaginaries in a Globalizing Age*, ed. Andreas Huyssen (Durham, NC: Duke University Press, 2008), 1–15.

19 Robinson, *Ordinary Cities*, 7.

20 Harvey, *The Condition of Postmodernity*; James Donald, "Metropolis: The City as Text," in *Social and Cultural Forms of Modernity*, eds. Robert Bocock and Kenneth Thompson (Cambridge: Polity Press, 1992), 418–61; and James Donald, *Imagining the Modern City* (London: Athlone, 1999).

21 For an overview, see: Zeynep Çelik, "New Approaches to the 'Non-Western City'," *Journal of the Society of Architectural Historians* 58, no. 3 (September 1999): 374–81.

22 Alev Çınar and Thomas Bender, "Introduction," in *Urban Imaginaries: Locating the Modern City*, eds. Alev Çınar and Thomas Bender (Minneapolis, MN: University of Minnesota Press, 2007), xxv; James Donald, "Metropolis: The City as Text," 427.

23 For an overview, see: Ahmet İçduygu, "From Nation Building to Globalization: An Account of the Past and Present in Recent Urban Studies in Turkey," *International Journal of Urban and Regional Research* 28, no. 4 (2004): 441–7.

24 Uğur Tanyeli, *Istanbul 1900–2000: Konutu ve Modernleşmeyi Metropolden Okumak* (İstanbul: Akın Nalça, 2004), 34.

25 İlhan Tekeli, *The Development of the Istanbul Metropolitan Area: Urban Administration and Planning* (Istanbul International Union of Local Authorities, Section for the Eastern Mediterranean and Middle East Region, 1994); İlhan Tekeli, *İstanbul'un Planlanmasının ve Gelişmesinin Öyküsü* (İstanbul: Tarih Vakfı Yurt Yayınları, 2013).

26 İlhan Tekeli, "The Story of Istanbul's Modernisation," *Architectural Design* 80, no. 1 (January/February 2010): 32–9. Also see "Cities in Modern Turkey," in *Istanbul: City of Intersections*, LSE Cities (November 2009), https://lsecities.net/media/objects/articles/cities-in-modern-turkey/en-gb/.

27 Erik Swyngedouw and Maria Kaïka, "The Making of 'Glocal' Urban Modernities: Exploring the Cracks in the Mirror," *City* 7, no. 1 (2003): 5–21.

28 Tahire Erman, "The Politics of Squatter (Gecekondu) Studies in Turkey: The Changing Representations of Rural Migrants in the Academic Discourse," *Urban Studies* 38, no. 7 (2001): 983–1002.

29 Çağlar Keyder, "A Brief History of Modern Istanbul," in *Turkey in the Modern World*, ed. Reşat Kasaba, vol. 4, *Cambridge History of Modern Turkey*, eds. Kate Fleet Kunt, Suraiya N. Faroqhi, and Reşat Kasaba (Cambridge: Cambridge University Press, 2008), 504–23.

30 Hatice Kurtuluş, ed., *İstanbul'da Kentsel Ayrışma: Mekansal Dönüşümde Farklı Boyutlar* (Istanbul: Bağlam Yayıncılık, 2005).

31 Didem Danış and Jean-François Pérouse, "Zenginliğin Mekânda Yeni Yansımaları: İstanbul'da güvenlikli siteler," *Toplum ve Bilim*, no. 104 (2005): 92–103; Ayfer Bartu Candan and Biray Kolluoğlu, "Emerging Spaces of Neoliberalism: A Gated Town and a Public Housing Project in Istanbul," *New Perspectives on Turkey*, no. 39 (2008): 5–46; and Şerife Geniş, "Producing Elite Localities: The Rise of Gated Communities in Istanbul," *Urban Studies* 44, no. 4 (2007): 771–98.

32 David Behar and Tolga İslam, eds., *İstanbul'da "Soylulaştırma": Eski Kentin Yeni Sahipleri* (Istanbul: Bilgi Üniversitesi Yayınları, 2006); Amy Mills, *Streets of Memory:*

Landscape, Tolerance, and National Identity in Istanbul (Athens, GA: University of Georgia Press, 2010); and C. Nil Uzun, "The Impact of Urban Renewal and Gentrification on Urban Fabric: Three Cases in Turkey," *Tijdschrift voor Economische en Sociale Geografie* 94, no. 3 (2003): 363–75.

33 Özlem Ünsal and Tuna Kuyucu, "Challenging the Neoliberal Urban Regime: Regeneration and Resistance in Başıbüyük and Tarlabaşı," in *Orienting Istanbul: Cultural Capital of Europe?*, eds. Deniz Göktürk, Levent Soysal, and Ipek Türeli (New York: Routledge, 2010), 51–70.

34 Dikmen Bezmez, "The Politics of Urban Waterfront Regeneration: The Case of the Golden Horn, Istanbul," *International Journal of Urban and Regional Research* 32, no. 4 (2008): 815–40.

35 Deniz Göktürk, Levent Soysal, and Ipek Türeli, eds., *Orienting Istanbul: Cultural Capital of Europe?* (London; New York: Routledge, 2010).

36 Ayfer Bartu Candan and Cenk Özbay, eds., *Yeni İstanbul Çalışmaları: Sınırlar, Mücadeleler, Açılımlar* (Istanbul: Metis, 2014); Ayşe Çavdar and Pelin Tan, eds., *İstanbul: Müstesna Şehrin İstisna Hali* (Istanbul: Sel Yayıncılık, 2013); Derya Özkan, ed., *Cool Istanbul: Urban Enclosures and Resistances* (Bielefeld: Transcript, 2015).

37 For an overview, see: Ahmet Yaşar, "Osmanlı Şehir Mekanları: Kahvehane Literatürü," *Türkiye Araştırmaları Literatür Dergisi* 3, no. 6 (2005): 237–56.

38 The transnational influences of doctoral-level graduate education have to be recognized for such new research avenues. Dissertations in Ottoman urban history that draw from visual representations include but are not limited to those by Çiğdem Kafesçioğlu, İffet Orbay, Kathryn Ann Ebel, and Shirine Hamadeh. For an overview, see: Kathryn Ebel, "Visual Sources for Urban History of the Ottoman Empire," *Türkiye Araştırmaları Literatür Dergisi* 3, no. 6 (2005): 457–86.

39 Tony Bennett, *The Birth of the Museum: History, Theory, Politics* (London: Routledge, 1995), 59–88.

40 Jean Baudrillard, "The Precession of Simulacra," in *Simulacra and Simulation*, trans. Sheila Glaser (Ann Arbor: University of Michigan Press, 1994), 1–42.

Bibliography

Ahıska, Meltem. "Occidentalism: The Historical Fantasy of the Modern." *The South Atlantic Quarterly* 102, no. 2/3 (2003): 351–79.

"AKM'yi yıkacağız, Taksim'e cami de yapacağız: Başbakan Erdoğan, Taksim'i yenileme çalışmalarında kararlı olduklarını yineledi." *Dünya*, June 2, 2013. Accessed June 23, 2013. http://www.dunya.com/akmyi-yikacagiz-taksime-cami-de-yapacagiz-193887h.htm.

Aktuğ, Erkan. "Hayal-et'ti, gerçek olacak!" *Hürriyet*, June 3, 2011. Accessed on January 10, 2017. http://www.radikal.com.tr/turkiye/hayal-etti-gercek-olacak-1051623/.

Arango, Tim. "Mosque Dream Seen at Heart of Turkey Protests." *New York Times*, June 23, 2013. Accessed June 23, 2013. http://www.nytimes.com/2013/06/24/world/europe/mosque-dream-seen-at-heart-of-turkey-protests.html?pagewanted=all.

Ashworth, Gregory J. *War and the City*. London; New York: Routledge, 1991.

Bartu Candan, Ayfer, and Biray Kolluoğlu. "Emerging Spaces of Neoliberalism: A Gated Town and a Public Housing Project in Istanbul." *New Perspectives on Turkey*, no. 39 (2008): 5–46.

Bartu Candan, Ayfer, and Cenk Özbay, eds. *Yeni İstanbul Çalışmaları: Sınırlar, Mücadeleler, Açılımlar*. Istanbul: Metis, 2014.

Baudrillard, Jean. *Simulacra and Simulation*, translated by Sheila Glaser, Ann Arbor: University of Michigan Press, 1994.

Behar, David, and Tolga İslam, eds. *İstanbul'da "Soylulaştırma": Eski Kentin Yeni Sahipleri*. Istanbul: Bilgi Üniversitesi Yayınları, 2006.

Bennett, Tony. *The Birth of the Museum: History, Theory, Politics*. London: Routledge, 1995.

Bezmez, Dikmen. "The Politics of Urban Waterfront Regeneration: The Case of the Golden Horn, Istanbul." *International Journal of Urban and Regional Research* 32, no. 4 (2008): 815–40.

Boym, Svetlana. *The Future of Nostalgia*. New York: Basic, 2001.

Çavdar, Ayşe, and Pelin Tan, eds. *İstanbul: Müstesna Şehrin İstisna Hali*. Istanbul: Sel Yayıncılık, 2013.

Çelik, Zeynep. "New Approaches to the 'Non-Western City'." *Journal of the Society of Architectural Historians* 58, no. 3 (September 1999): 374–81.

Çınar, Alev, and Thomas Bender, eds. *Urban Imaginaries: Locating the Modern City*. Minneapolis: University of Minnesota Press, 2007.

Danış, Didem, and Jean-François Pérouse. "Zenginliğin Mekânda Yeni Yansımaları: İstanbul'da güvenlikli siteler." *Toplum ve Bilim*, no. 104 (2005): 92–103.

Donald, James. "Metropolis: The City as Text." In *Social and Cultural Forms of Modernity*, edited by Robert Bocock and Kenneth Thompson, 418–61. Cambridge: Polity Press, 1992.

—. *Imagining the Modern City*. London: Athlone, 1999.

Ebel, Kathryn. "Visual Sources for Urban History of the Ottoman Empire." *Türkiye Araştırmaları Literatür Dergisi* 3, no. 6 (2005): 457–86.

Erman, Tahire. "The Politics of Squatter (Gecekondu) Studies in Turkey: The Changing Representations of Rural Migrants in the Academic Discourse." *Urban Studies* 38, no. 7 (2001): 983–1002.

Gaonkar, Dilip Parameshwar, ed. *Alternative Modernities*. Durham, NC: Duke University Press, 2001.

Geniş, Şerife. "Producing Elite Localities: The Rise of Gated Communities in Istanbul." *Urban Studies* 44, no. 4 (2007): 771–98.

Göktürk, Deniz, Levent Soysal, and Ipek Türeli, eds. *Orienting Istanbul: Cultural Capital of Europe?* London; New York: Routledge, 2010.

Harvey, David. *The Condition of Postmodernity: An Enquiry into the Origins of Cultural Change*. Cambridge, MA: Blackwell, 1990.

Huyssen, Andreas, ed. *Other Cities, Other Worlds: Urban Imaginaries in a Globalizing Age*. Durham, NC: Duke University Press, 2008.

İçduygu, Ahmet. "From Nation Building to Globalization: An Account of the Past and Present in Recent Urban Studies in Turkey." *International Journal of Urban and Regional Research* 28, no. 4 (2004): 441–7.

Ipsen, Detlev. "The Socio-Spatial Conditions of the Open-City: A Theoretical Sketch." *International Journal of Urban and Regional Research* 29, no. 3 (2005): 644–53.

"İstanbul 2010'a ilginç reklam kampanyası." *Hürriyet*, December 4, 2009. Accessed February 28, 2010. http://www.hurriyet.com.tr/gundem/13103553_p.asp.

Keyder, Çağlar. "A Brief History of Modern Istanbul." In *Turkey in the Modern World* edited by Reşat Kasaba. Vol. 4. *Cambridge History of Modern Turkey*, edited by Kate Fleet Kunt, Suraiya N. Faroqhi, and Reşat Kasaba, 504–23. Cambridge: Cambridge University Press, 2008.

—, ed. *Istanbul: Between the Global and the Local*. Lanham, MD: Rowman & Littlefield, 1999.

—. "How to Sell Istanbul?" *İstanbul*, no. 3 (1992): 80–85.

Kozar, Cem. "Hayal-Et Yapılar Sergisi'nde Taksim Kışlası." *Arkitera.com*, February 29, 2012. Accessed January 10, 2017. http://www.arkitera.com/gorus/268/hayal-et-yapilar-sergisi-nde-taksim-kislasi.

Kurtuluş, Hatice, ed. *İstanbul'da Kentsel Ayrışma: Mekansal Dönüşümde Farklı Boyutlar*. Istanbul: Bağlam Yayıncılık, 2005.

Mills, Amy. *Streets of Memory: Landscape, Tolerance, and National Identity in Istanbul*. Athens, GA: University of Georgia Press, 2010.

Özbay, Ferhunde. "İstanbul'da 1950 sonrası Nüfus Dinamikleri." In *Eski İstanbullular Yeni İstanbullular*, edited by Murat Güvenç, 54–77. Istanbul: Osmanlı Bankası Arşiv ve Araştırma Merkezi, 2009.

Özkan, Derya, ed. *Cool Istanbul: Urban Enclosures and Resistances*. Bielefeld: Transcript, 2015.

Özyürek, Esra. *Nostalgia for the Modern: State Secularism and Everyday Politics in Turkey*. Durham, NC: Duke University Press, 2006.

Rieniets, Tim, Jennifer Sigler, and Kees Christiaanse, eds. *Open City: Designing Coexistence*. Amsterdam: SUN Architecture, 2009.

Robinson, Jennifer. *Ordinary Cities: Between Modernity and Development*. London: Routledge, 2006.

Sennett, Richard. "The Open City." In *The Endless City*, edited by Ricky Burdett and Dejan Sudjic, 290–7. London: Phaidon, 2008.

Swyngedouw, Erik, and Maria Kaïka. "The Making of 'Glocal' Urban Modernities: Exploring the Cracks in the Mirror." *City* 7, no. 1 (2003): 5–21.

Tanyeli, Uğur. *Istanbul 1900–2000: Konutu ve Modernleşmeyi Metropolden Okumak*. İstanbul: Akın Nalça, 2004.

Tekeli, İlhan. *The Development of the Istanbul Metropolitan Area: Urban Administration and Planning*. Istanbul International Union of Local Authorities, Section for the Eastern Mediterranean and Middle East Region, 1994.

—. "Cities in Modern Turkey." In *Istanbul: City of Intersections*. LSE Cities, November 2009. https://lsecities.net/media/objects/articles/cities-in-modern-turkey/en-gb/.

—. "The Story of Istanbul's Modernisation." *Architectural Design* 80, no. 1 (January/February 2010): 32–9.

—. *İstanbul'un Planlanmasının ve Gelişmesinin Öyküsü*. Istanbul: Tarih Vakfı Yurt Yayınları, 2013.

Ünsal, Özlem, and Tuna Kuyucu, "Challenging the Neoliberal Urban Regime: Regeneration and Resistance in Başıbüyük and Tarlabaşı." In *Orienting Istanbul: Cultural Capital of Europe?*, edited by Deniz Göktürk, Levent Soysal, and Ipek Türeli, 51–70. New York: Routledge, 2010.

Uzun, C. Nil. "The Impact of Urban Renewal and Gentrification on Urban Fabric: Three Cases in Turkey." *Tijdschrift voor Economische en Sociale Geografie* 94, no. 3 (2003): 363–75.

Yaşar, Ahmet. "Osmanlı Şehir Mekanları: Kahvehane Literatürü." *Türkiye Araştırmaları Literatür Dergisi* 3, no. 6 (2005): 237–56.

2 Picturing "Old Istanbul"

At the turn of the millennium, a sense of loss seemed to permeate public discussions regarding the future of Istanbul, while the city seemed more modern and connected than ever. As the city surfaced in global networks, it appeared to drown in nostalgia. Nostalgia, here, denotes a collective feeling, "a longing for a home that no longer exists or has never existed."[1] This yearning arises from dissatisfaction with the present. It may be for a place, but it can also be a longing for a different time. One of the ways nostalgia manifests itself in a city is through the imagining of a cosmopolitan and more colorful past. Visual and literary depictions become important sites through which to imagine and consume bygone times. Characteristically, Istanbul's bookstores stocked their shelves with works depicting the city's past. Among the wide range of publications on "Old Istanbul" for sale, those with reproductions of photographs by Ara Güler became especially popular. Güler's melancholic black-and-white pictures of Istanbul in the 1950s and 1960s experienced a renaissance in the early 1990s and have since become synonymous with a particular form of longing for Istanbul's past.

Most of Güler's best-known photographs are from the beginning of his career, when he roamed the streets as a photojournalist. In the 1950s, major newspapers and magazines in Istanbul added photographers to their staffs and started carrying serial articles and dedicated photo "corners" portraying rapid urban change and the impact of poor rural-to-urban migrants on the city. Turkey was being integrated into the postwar economic order at the time, and major public-works projects were underway in Istanbul. Just as new boulevards cut through the city in response to the demands of urbanization, photographs in print media cut through the talk on the printed page and opened see-for-yourself windows onto the city.

What is it in these black-and-white images of the urban poor and working class that today lend them so powerfully to the sense of nostalgia? Informed by existing discussions of nostalgia, cosmopolitanism, and media (or prosthetic) memories, this chapter addresses how Güler's pictures, originally commissioned by journalism outlets in response to very specific changes in the city, have resurfaced in art galleries, on the pages of coffee-table books, and in other venues of cosmopolitan nostalgia. I examine how meanings ascribed to images may be transformed as conditions of reception and audiences change over time.

Photographs recalled from the archive

Ara Güler is today the favorite and quasi-official photographer of Istanbul. Newspaper and television reports in the spring of 2008 claimed Güler was not pleased that one of the most recent exhibitions of his Istanbul work, "İstanbul'da Alınteri" (Sweat [of Labor] in Istanbul), had been mounted in the Taksim subway station (see Figure 2.1). Reportedly, Güler had supplied the photographs but was not informed as to how they would be exhibited; he thus claimed that the photographs in the show, organized jointly by the Istanbul Chamber of Commerce and the Istanbul Metropolitan Municipality, needed to be exhibited in an art gallery and that ordinary metro passengers who walked by his photographs in the station would not understand them.[2] The irony, of course, is that most of these photographs had originally been taken for newspapers and illustrated magazines, not for exhibitions in galleries.

This irony is connected to debates on the valorization of photography as an art of the original and the photographer as artist in the modernist sense. The move of documentary, news, and even scientific (e.g. topographic views) photography into the exhibition and the museum in the 1980s was problematized early on by art critics such as Christopher Phillips, Rosalind Krauss, and Abigail Solomon-Godeau.[3] Douglas Crimp observes that photographs that may have previously been illustrations in books and organized (in the space of the library) by topic came to be classified by the "artists" who made them, leading, for instance, to

Figure 2.1 Opening of the exhibition "İstanbul'da Alınteri" (Sweat [of Labor] in Istanbul) mounted in the Taksim subway station. Güler poses with Istanbul's Mayor Kadir Topbaş. The board to the left is plastered with a well-known photograph by Güler, "Eminönü after the 1959 demolition work," which captures a moment of urban renewal by overlaying the horse-drawn cart with a car on a boulevard in the middle ground and a mosque-marked skyline in the background.

Istanbul Metropolitan Municipality website, accessed January 5, 2017, www.ibb.gov.tr/tr-TR/SiteImages/Haber/mart2008/ara_sergi27.jpg.

what would have formerly been characterized as "urban poverty" to recirculate as Jacob Riis or Lewis Hine shots.[4] Echoing such developments, Güler's photographs of poor urban working classes taken in the 1950s and 1960s were recalled from the archive, detached from their original presentation in newspapers and illustrated magazines—usually to accompany texts—and moved into exhibition contexts. The recirculation (and popularity) of these photographs since the 1990s meant a small section of his work came to be characterized as his artistic signature.

Güler's career as a professional photographer took off in the context of expanding print media and in relation to dramatic physical and social transformations in the city during the 1950s due to urban renewal and rapid population growth from internal migration. However, since the early 1990s the reframing of Güler's work through a selection of his photographs from his earlier work that document change on the streets of Istanbul has a much different purpose. This reframing intends to promote a popularized longing for the multiethnic, multi-religious past of the city, the loss of which the photographer also laments and explicitly states in interviews. Güler's pictures are powerful today because they preserve the city in a way that is particularly useful to nostalgic accounts of the city's cosmopolitan past. The work is significant not only because of the photographs' content but also due to its recirculation in a variety of media, ranging from books and exhibitions to the built environment.

The revival of Güler's work made his name synonymous with an urban nostalgia that was amplified into the realm of popular consumption (of visual cultural products) in the early 1990s. This corresponded with the beginnings of an official government effort to transform Istanbul into a "capital of culture"—reflective of the global turn from managerial to entrepreneurial city governance and the concomitant efforts to use "culture" to market cities.[5] One effect was an efflorescence of publications regarding the past of the city. As early as 1992, arts and culture weekly *Milliyet Sanat* featured on its cover (no. 286) the title "Istanbul that takes refuge in books" with a visual that portrayed a person equipped with a camera zooming in on a photograph of an old wooden house found in a book on the history of Istanbul (see Figure 2.2).[6] Such a visual acknowledges the newfound popularity of books on Istanbul (the book is rendered larger than the individual looking at it), our technologically mediated access to that history (which requires, as depicted, a printed book and a camera), and our distance from it (the camera needs a tele-lens to see into the metaphorical distance of time past). It is not necessarily that Istanbul takes refuge in books, as the cover title suggests; it is the individual who takes refuge in the mediated images of the city.

It is possible to date the revival of Güler's signature work to 1992, when Turkey's Ministry of Culture and Tourism declared the following year to be the "Year of Istanbul" and recruited the History Foundation to manage and organize the celebrations. One of the self-designated missions of this foundation, a then newly established civil society organization composed of prominent Turkish social scientists and historians, was to challenge official historical narratives by focusing on the everyday social and cultural history of Istanbul from a minor history perspective.[7] As part of the celebrations, it organized Güler's first Istanbul

Figure 2.2 "Istanbul that takes refuge in books." Cover of arts and culture weekly *Milliyet Sanat* illustrates a dossier on the resurgence of books on Istanbul's history in reaction to rampant, uncontrolled urbanization.

Milliyet Sanat, no. 286 (April 15, 1992), 12–19. With contributions from Abdullah Kuran, Enis Batur, and Alpay Kabacalı, and blurbs from Murat Belge, Zeynep Avcı, Hulki Altunç, Ahmet Turhan Altıner, and Mario Levi.

exhibition, "Istanbul, an Endless Reportage," his twenty-second solo show. A coffee-table catalogue, *A Photographical Sketch on Lost Istanbul*, subsequently presented Güler's selected Istanbul work in print in 1994.[8] Many other photographic exhibitions and books followed in this vein, all with the intention of reproducing the "lost" Istanbul Güler had captured. That exhibitions of Güler's photographs represented Turkey at the nine-month "Turkish Cultural Season" in France in 2009 or the 2015 "Europalia Arts Festival" demonstrate that he remains the photographer of choice for representing Turkey officially.[9]

Photographs on exhibit, in and beyond the gallery

Güler's well known Istanbul photographs have been reproduced in many forms, and their inclusion in exhibitions, magazines, and books has provided varied opportunities for imagining Old Istanbul. Some commercial ventures have gone even further, advertising physical environments in which customers may immerse themselves in Güler's work. Two striking examples are the Ara Café, located in a building owned by Güler in the district of Beyoğlu (a building that also houses his studio), and Point Hotel, a boutique hotel in the district of Talimhane, where Güler was born. These two commercial spaces have been likened to "museums" in promotional reports.[10]

It was Güler who chose the black-and-white photographs from the 1950s and 1960s to be placed in these spaces.[11] Photographers visiting Istanbul on the occasion of the Magnum exhibition, "Turkey by Magnum" (Istanbul Modern Museum, 2007), were hosted at the sponsoring Point Hotel and allegedly they too marveled at the museum-like feeling of the hotel's decoration concept, in which rooms, restaurant, lobby, and other common spaces are adorned with, according to the hotel, 232 of Güler's works.[12] Güler further helped promote the hotel by posing with his camera in front of his photographs on the walls.[13]

At the nearby Ara Café, customers eat on paper placemats on which Güler's well-known Istanbul photographs are printed, while they can study large-format reproductions of his images on the walls. This gives the café customers the possibility of imagining that at any minute the photographer himself might come down from his studio to join them. Indeed, he is often spotted there, and he has given many of his interviews in this space.[14]

The exhibitionary aspect of these two spaces may be linked to the growth of new private museums and alternative exhibition spaces in the city in the past two decades. Looking beyond Istanbul, however, it is difficult not to notice that the use of black-and-white or monochrome prints in cafés, bars and similar commercial collective spaces emerged in the 1990s as a "signature of designer chic" in metropolitan centers.[15] The doubling of these two commercial spaces, Ara Café and Point Hotel, as exhibition spaces is indicative of how the reframing of Güler's images is part of a broader, transnational cultural and commercial production.

As with most constructions of memory, this process of evoking Old Istanbul entails remembering and forgetting in a dialectical relationship, where memory and history are entangled rather than oppositional, as Pierre Nora argued in his famous work on *lieux de mémoire*.[16] Photographs, however, are controversial and complex *lieux de mémoire*. Photography, popularly conceived as an act of remembrance, is "at odds" with and "blocks memory," according to some of photography's best-known critics, such as Siegfried Kracauer and Roland Barthes.[17] In contrast, historian of photography Geoffrey Batchen has suggested that photographs are not necessarily produced to bring the past to the present but rather to situate the self in relation to an unknown future.[18] In the context of Istanbul, the recalling of old photographs from the archive serves a similar purpose: In a city under the strain of economic neoliberalization, an older or other time and place can be imagined as a *belle époque*, a time when citizens were imagined to be more civilized and tolerant toward each other; a vaguely defined temporality. Güler's photographs have thus been instrumental in imagining a more harmonious past as the basis for an "open city" in the future.

This process of curating the past for the needs of the present and for imagining the future is not exclusive to Istanbul. Many cities are undergoing similar processes. The notion of Old Istanbul ties in well with official agendas of city marketing, as newly configured "nostalgic" spaces organically become linked to the pursuit of middle class dreams and consumption patterns. The flourishing of the market economy has created a new dynamic concerning the past—not only in Istanbul but also in other cities competing for a place on the map of global cities. Municipal governments attempt to capitalize on the past by revalorizing what frequently are empire-era reminders in their bid to compete in the globalized economy. Furthermore, the appreciation of "Old Istanbul" has emerged as a means of group distinction as well as displaced social criticism. Among other media, photography has proved particularly instrumental in these processes because of its indexicality—or truth claimed due to the physical relationship between object photographed and photograph.

Cosmopolitan nostalgia in search of the "Istanbulite"

The photographs feed, not any, but cosmopolitan nostalgia. They feed a yearning for a time when Istanbul was less homogenized and culturally richer. In the past decades, there has been a plethora of publications on cosmopolitanism or "cosmopolitics" in English-language academic literature. In this work, the term cosmopolitanism refers to the ideal of an international system based on democratic principles and the rule of law or to global citizenship, both of which are positioned against the specter of nationalism or fundamentalism.[19] For instance, according to Pheng Cheah, "cosmopolitanism is the obvious choice as an intellectual ethic or political project that can better express or embody genuine universalism."[20]

There seems to be a general popular and academic understanding within Middle East Studies that certain forms of cosmopolitanism existed at the time of the Ottoman Empire, but they were later hindered by modernization. The introduction to a volume on cosmopolitanism in the region argues that

> during the Ottoman period, the Middle East was an open, undefined territory in which groups of different religious and ethnic backgrounds intermingled and exchanged ideas and lifestyles. Cosmopolitanist cities—Alexandria, Istanbul and Beirut—formed free havens for cultural exchange. No definite and rigid boundaries had been drawn, and the state did not yet exert its power of standardization or impose its norms on its citizens.[21]

In such accounts, it is the formation of the nation-state that brought an end to cosmopolitan society.

Cosmopolitanism is not a conceptual framework imposed from without: Local scholars have also used, discussed, and criticized it.[22] The term *cosmopolitan* is used in the region and in Turkey in several ways—in reference to individuals, places, cities, and milieus. These meanings may overlap but not necessarily coincide.[23] The term is sometimes used interchangeably, in local discussions, with *multiculturalism* (*çok kültürlülük*) to refer to the empire-era coexistence of a multiplicity of cultures despite hierarchies and boundaries between them. However, their frame of reference is quite different: The objective of *multiculturalism* is "to satisfy ethnic and cultural minorities within the state," while *cosmopolitanism* refers to a world citizenship that transcends the borders of the nation-state.[24] As Benton Jay Komins points out, Istanbul's cosmopolitanism is "depopulated"; that is, the non-Muslim minority groups central to this imagination are mostly displaced.[25] The cosmopolitan label is frequently used in Turkey as a reaction to the perceived cultural homogeneity of the present.

Discussions on cosmopolitanism in regards to Istanbul have been coupled with discussions on "Istanbuliteness"—who is an Istanbulite and who is not—demonstrating a characteristic mediation of class conflict via culture.[26] "Istanbulite" is a problematic term; indeed, sociologist Ayşe Öncü qualifies it as a "myth." In reference to the Istanbul of the 1990s, she explains more precisely that "in a metropolis of numerous and fluctuating plurality of cultural hierarchies, the word 'Istanbullu' stands guard over the boundary between high and popular culture."[27] Yearnings for the cosmopolitan past have frequently served as a way to cover up contempt for the poor among certain advantaged and privileged groups and as a central theme of public discourse on the city produced by the latter.

In the Ara Güler documentary of 1998 by Nebil Özgentürk, the narrator suggests, rather uncannily, that Istanbul should thank Güler because his photographs enable it to maintain its "innocence." It is important to dwell on the choice of word here. "Innocence" implies a good time that once was, and yet implies later corruption. A complex chain of associations is thus at work. In some narratives, the Republic is held responsible for homogenizing a once multicultural society; in others, the mass migration of peasant Turks from Anatolia is assigned responsibility. In a typical newspaper interview (published in 1997, in the *New York Times*), the photographer presents a demonstration of the "chain of associations" that connect the (loss of) cosmopolitanism to rural-to-urban migration and urban poverty.[28] He first claims he knows Istanbul because he grew up in the city; later, he laments that the aesthetic uniqueness or "poetry" of Istanbul is gone. He reasons, then, it is because of rural-to-urban migrants. They have "overrun" the city, constituting, in his estimate, twelve of its thirteen million population (at the time). He concludes that it is because of the migrants that non-Muslim minorities have left. In many such populist, as well as some scholarly, accounts, which are effectively "public transcripts" or required performances, the continued arrival of rural-to-urban migrants is associated with the exodus of non-Muslim minorities and the transfer of the latter's properties to Muslim Turks, a process referred to as the "Turkification" of the city. While both processes have to do with evolving state policies, no connection of causality exists between the arrival of rural-to-urban migrants from Anatolia and the departure of the city's non-Muslim residents. Such a view also attributes a romantic, nostalgic quality to the city prior to the 1950s, when much of the official policy of demographic homogenization targeting non-Muslims (e.g. special taxation, population exchanges, favoring the employment of Turkish-speaking Muslims over others, campaigns for speaking Turkish, clothing reforms) was enacted.[29] Writing in 2002, Rıfat Bali explains most vividly:

> The wave of the "Old Istanbul" nostalgia awakened an "urbanite consciousness." "To be an Istanbulite" was deemed a privilege ... In this environment, it became ordinary to come across journalistic series, interviews, and books on Istanbul. Prominent personalities who were consulted in these journalistic series reacted to the "invasion" of Istanbul by those from Anatolian backgrounds.[30]

The discussions on Istanbulite consciousness, which were alive well into the 2000s, drew from a well-recognized repertoire of expressions that included not only "overrun" and "invasion" but also "cultural degeneration" and "provincialization." These expressions had emerged nearly half a century earlier in response to the arrival of Anatolian rural-to-urban migrants in the city *en masse*, but by the 1990s internal migration had ceased to be the main source of the city's demographic growth.

These associations were, in fact, triggered by a phase of government-led urban renewal that commenced in the latter part of the 1980s and was intended to open the city to global capital. However, they quickly metamorphosed into a civilizing mission, adopted by associations and organizations of civil society and popularized by public intellectuals—individuals with access to the media. To this way of thinking, Istanbulite consciousness was located in the city's multicultural heritage, not in its current diversity.

With the rise of "Islamists" to power in 2003, the currency of these associations declined among secular liberals. However, Islamism-associated groups and especially local administrations subsequently cultivated new versions of them, a development attesting to the appeal of nostalgic urban modernity across the political spectrum. For example Istanbul Metropolitan Municipality under the AKP conducted a large-scale survey of "Istanbuliteness" (Kentim Istanbul) in 2003 and decorated the city's billboards with posters to promote Istanbulite consciousness.[31] It further sponsored many activities and publications, e.g. the coffee table book *I am Istanbul* (2007), pertaining to the multiethnic, multi-religious character of the city, with a misleading over-representation of the city's dwindling non-Muslim residents.

To be precise, then, the use of the term cosmopolitan in the public culture of Istanbul bears no reference to "citizens of the world," nor does it seek to turn Istanbul into a "city of refuge."[32] Instead (to use David Lowenthal's term), it is used loosely to refer to the Istanbul of the past as a "foreign country," where non-Muslim minorities and Muslim populations enjoyed mutual hospitality.[33] Clearly drawn boundaries and conflicts between different constituencies in the past do not enter present-day narratives of cosmopolitanism. It is this repainted image that seeks to reframe Turkey's future.

The original context of Güler's work

It is important to understand the original context of Güler's photographs to fully comprehend how they have been reframed in the present. This awareness may also provide a more nuanced understanding of the general workings of nostalgia in urban modernity. Only a handful of Güler's photographs have been reproduced from the total body of his work, which is said to consist of nearly 800,000 images. The reproduced photographs are mostly from the beginning of his career, in the 1950s and 1960s. Moreover, these continually reproduced photos do not refer to minorities or to social conflict. In other words, although they depict immense poverty, they do not lend themselves to pedagogical or reformist readings. Still, as I discuss in this section, Güler's oeuvre evolved necessarily in conversation with a broader range of visual and journalistic approaches to depicting urban poverty.

Bourgeois anxieties about the above-quoted "villagers from Anatolia" may be traced to the 1950s, when unprecedented numbers of poor migrants began arriving in the rapidly industrializing city. Economically driven internal migration had a profound effect on the shape and culture of Istanbul throughout the latter half of the twentieth century. It led to a verbose discourse about "integrating migrants" and anxieties about the "provincialization" of the city that paralleled debates in migrant labor—receiving countries in Europe.[34] The massive population increase came as a shock not only to those struggling to adapt to a new urban life, but also to the more established urban population.

The city's modernizing infrastructure, especially new boulevards built in the latter part of the 1950s, highlighted social and cultural differences between different social groups. The planning of Istanbul by Henri Prost, which guided the infrastructural modernization of the city until the 1960s, has been examined by Cânâ Bilsel and Pierre Pinon (2010), and Murat Gül (2004, 2009). The review of journalistic coverage shows that the "primal modern scenes," to use Marshall Berman's words, that arose from the concrete everyday life of Istanbul during this stage of urban modernity were quite different from those other modernizing metropoles, including those in Ankara (in the early Republican era, as discussed by Gülsüm Baydar, for example).[35] In Istanbul, physical modernization and the expansion of media outlets occurred simultaneously. Together, these two development forces created the stage on which migrants were exhibited for the gaze of the city's middle-class residents.

While migrants arrived *en masse* on buses, trains, and ships, the arrival of crowded migrant boats made headlines with dramatic accompanying photographs of packed decks. Tophane, the site where luxury cruise ships dock to bring elite tourists to Istanbul today, was their landing point. One of the first to document these "migrant boats" was Yaşar Kemal, one of Turkey's most well-known novelists, who also earlier pursued a career in journalism. Kemal's early experiences among peasants and workers made him a devoted defender of the underprivileged and a sustained critic of contemporary bourgeois culture.

Kemal took the Black Sea boat from Samsun to Istanbul in 1952 to produce one of his first stories for the *Cumhuriyet* newspaper.[36] In a touching account of poverty and misery, Kemal narrated how the cargo boats, which were incredibly slow and without appropriate accommodations, picked up migrants from port cities all along the north coast of Turkey. Their journey to Istanbul lasted for nearly ten days. Hundreds of men, women, and children—all strangers—had to sit pressed up against each other, and if they were accommodated on deck, they had to stand and sleep in the rain. Kemal was appalled by the congestion and filth on these boats. Once they arrived—and three full boats arrived per week—the migrants on them would pour into Tophane (see Figure 2.3). There they would begin to look for jobs in the cafés and on the streets, and sleep wherever they could—if only under the shade of a tree or on the side of the road. The city was flooded with newcomers.

The arrival of migrants from the early 1950s onwards was viewed as an extension of the existing housing problem in the city, rather than a labor problem.[37] Thus, the initial public and administrative reaction was to ask the migrants to leave. Reportedly, the governor's office tried to send migrants back to their place

Figure 2.3 A boat carrying rural-to-urban migrants to Istanbul, similar to the one Yaşar Kemal takes for his research for his *Cumhuriyet* serial interview in 1952.

Hürriyet, July 29, 1955, 1.

of origin by purchasing five hundred return tickets every month, but this measure proved far from effective.[38] Meanwhile, the physical environments that migrants built for themselves were so primitive that social and health problems were inevitable. A four-day "survey" in *Cumhuriyet* in 1951 alleged that "the unfortunate fate of those who migrated from Anatolia" was criminality.[39]

Nevertheless, by 1960 the squatter settlements had acquired a sort of sensational appeal. Influenced by ideas of socialism in politics and social realism in art, groups of journalists began conducting serial interviews with the migrants rather reproducing the official point of view. In this regard, they were far ahead of social scientists, who studied migration as a problem of "rural disintegration" and were just turning their attention to the experiences of the migrant population of the city.[40] Still, most journalistic reports overlooked the differences between the migrants and tended to concentrate on common problems with the migrants' crudely made physical environments and living conditions.

This new contradiction in the character of the city—the increasing display of poverty compared to new displays of wealth or the presence of cattle as well as new American cars on the boulevards—resulted in a complex set of anxieties. These anxieties had to do with bourgeois fears of invasion, contamination, and criminality, as well as with the uncertainty of class identity—that is, an inability to tell who was an authentic Istanbulite. Some of the newcomers had access to the consumption practices enjoyed by more established members of the middle class, while others, who were even better off, had greater access to certain luxuries.

Gender imbalance (the first migrants to the cities and those looking for jobs on the streets were mostly men) played into fears of assault and criminality. For example, Selma Emiroğlu's "Bayan Istanbulu" (Ms. Istanbulite), the heroine of the cartoon series, is worried about a man she encounters on a side street, who is dressed clearly in a poorer person's attire; she fantasizes about a potential attack but is relieved when he abuses her only verbally: "I was afraid for no reason," she explains, "he was after all a good man and did not cause too much harm" (see Figure 2.4). The cartoonist acknowledges women's particular experiences but also parodies the hyperbole of such fears.

Two other seminal cartoons, the likes of which have appeared repeatedly in the press since the 1950s, capture the main threads of anxieties: invasion and uncertainty of appearances. They had, of course, the additional intention of provoking and influencing public opinion. A cartoon by Turhan Selçuk from 1958 ridicules both the notion of population "explosion" and the proposal to control entry into Istanbul (see Figure 2.5). The caption—"Keep on going gentlemen, there is room at the front"—refers to a colloquialism used in crowded public buses by those who hopped on through the back door without paying their tickets. This cartoon suggests through its caption and image that Istanbul had become an open city. The notion of the open city reappeared in various forms during the 1950s and 1960s. During those years, the captions to now-faded front-page photographs, editorials, and cartoons suggested on a daily basis that Istanbul had become an open city, with "all sorts of people on the streets," both a cynical reference to the

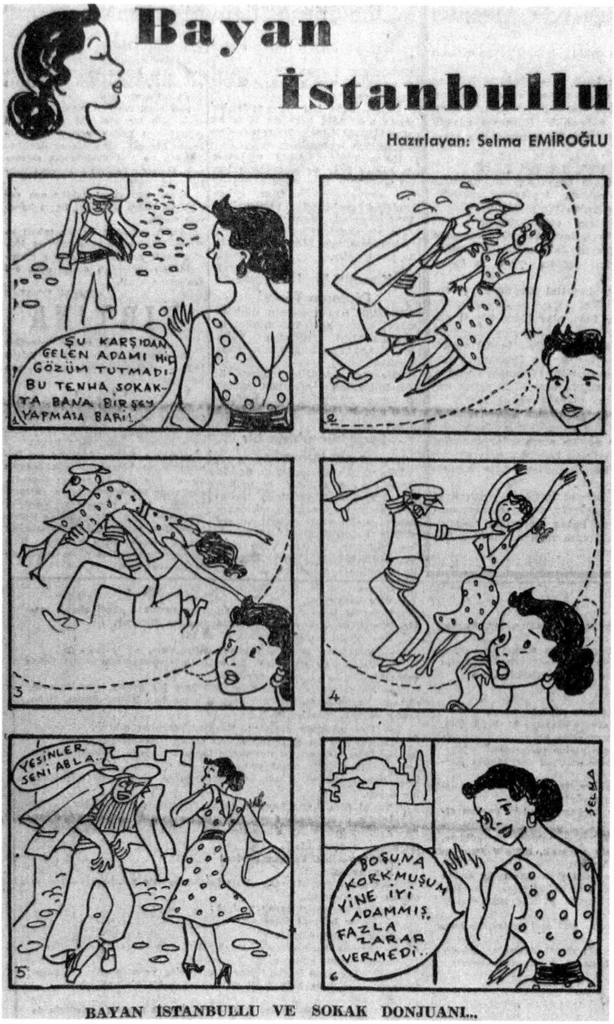

Figure 2.4 "Ms. Istanbulite" cartoon self-reflexively mocks fear of being harassed by (rural-to-urban migrant) men from an urbanite woman's perspective.

Selma Emiroğlu, "Bayan Istanbullu," *Yeni İstanbul*, June, 30, 1957, 6.

new visibility of the urban poor and an expression of appreciation for the city's new look and feel.

The cartoon by Cafer Zorlu responds to the general shock expressed in newspapers over mass migration (see Figure 2.6). Istanbul is represented here in the background by a minaret and a cluster of small-town, small-scale houses. A split person appears in the foreground: one half suited as an Istanbulite, the other dressed as a peasant in work clothes with a lengthy moustache. This cartoon contrasts

Figure 2.5 Istanbul rendered as an "open city," one that has abandoned defensive efforts and allowed invaders to march in. The caption reads: "Keep going gentlemen, there is room at the back."

Cartoon by Turhan Selçuk, *Kent, Konut ve Yerleşim Üzerine Karikatürler = the City, Homes & Settlements in Caricature (1908–1995)*, ed. Turgut Çeviker (Istanbul: Devlet Güzel Sanatlar Galerisi, 1996), 44–45 (Reproduced from *Milliyet* 1959).

to early Republican representations of urbanites and peasants. Sibel Bozdoğan points out that in many of the posters advertising government-produced consumer goods during the 1930s, the peasant and the urbanite existed in "harmonious coexistence … with the products of industry offering fulfillment of their respective aspirations in life—parallel but never conflicting."[41] In contrast, Zorlu's split figure acknowledges widespread concern that Istanbul has been "invaded." The caption, "Half of Istanbul is of rural origin. (Newspapers)," emphasizes the role played by newspapers in reporting and shaping public opinion on the issue. The cartoonist may or may not have endorsed the view that demographic transformation is negative, but he has opted to express and analyze the contemporary anxiety revealed in newspaper headlines. The plight of migrants is clearly not the main issue here. Instead, the split appearance of the man takes center stage. Indeed, the cartoon can be read as a commentary on the uncertainty of appearances: Is the man really an Istanbulite, or is he a pretender?

In contrast to Yaşar Kemal's humanism and the self-reflexive satire in cartoons discussed above, much of the journalistic visual presentation of the city's transformation relied on terse, didactic messages. A typical use of street photography at the time was daily "corners" on the front pages of leading newspapers. These tended to juxtapose the urban poor and migrants with modern spaces. Usually no larger than one third of the width of the page and with only a simple caption underneath, this strategic framing accentuated both the backwardness of the migrants and their potential to become "modern." Typical of these is a photograph (by Sökmen Baykara) showing two peasant men walking a herd of cattle on a brand-new avenue right in front of the recently opened building of the Istanbul

Figure 2.6 The cartoon is both an allusion to the perceived "ruralisation" of the city, as depicted in contemporary newspapers, and to the uncertainty of appearances. The caption reads: "Half of Istanbul is of rural origin. (Newspapers)."

Cartoon by Cafer Zorlu, *Kent, Konut ve Yerleşim Üzerine Karikatürler = the City, Homes & Settlements in Caricature (1908–1995)*, ed. Turgut Çeviker (Istanbul: Devlet Güzel Sanatlar Galerisi, 1996), 48 (Reproduced from *Akbaba*, 1961).

Municipality. The cattle, signifying traditional ways, block the roads, which have been paved for modern vehicles. The caption reads:

> Right Under [their] Nose: Administrators inside the Municipal Palace oversee order in the city, citizen rights and public health, decisions on social issues. As seen in the picture, cows are being herded slowly by their owners in front of the Municipality.[42]

Together with its caption, the photograph suggests that such sights ought not to be allowed by the authorities. The cattle (or traditional ways) should not block roads paved for modern vehicles. At other times, sheep, horse carts, peasants, and squatter settlements were similarly depicted in front of wide-open asphalt boulevards and modernist buildings. "Is it possible to see in any other big city other than Istanbul that claims to be European [when] herds of cattle pass through the largest square of the city in broad daylight?" one caption asked.[43] This kind of representation supported the hegemonic belief that an inefficient traditional society impeded the progress of modernization—a mystification that neglected to address the economic inequality causing such scenes in the first place.

The social and physical context in which Güler's photographs of Istanbul were originally produced was one of rapid urbanization, infrastructural modernization, and an opening of the city's economy and culture. The rise of an urban consumer culture, the development of mass-media outlets, and the construction of new public spaces enabled new encounters and vistas of the city. In particular, the opening of new boulevards through the built fabric of the city during the 1950s, along with the rise of mass-circulation newspapers and magazines, made the simultaneous arrival of waves of rural migrants visible and heightened public anxieties about urban change. Visual depictions of the city in print journalism, in turn, guided public discussions on urban transformation and helped shape public opinion.

Print windows onto the city

Despite early encounters with various media of "technological reproducibility"—photography, printing, and cinema—"mass" demand for these outlets only picked up during the post–World War II period. Existing literature attributes this belated increase in circulation to the political democratization that swept through Turkey after the multiparty elections of 1950.[44] However, it may also be argued that this resulted from a delay in consumer-reader demand. The rise of the mass-print press in Istanbul and the urbanization of the city were very much related. Mass-circulation dailies and magazines translated the everyday experiences of the city into images for consumption. And, as part of this effort, the use of interviews and photographs reflected a desire for spontaneity and simultaneity on the part of publications that would also enable readers to identify with journalists and with other readers as a "community" of Istanbulites.

The *Hürriyet* and *Yeni İstanbul* newspapers, along with *Hayat* magazine, became pioneers of this trend by prioritizing the use of photographs and hiring photographers onto their staffs. *Hürriyet*, published from May 1, 1948 through the present, has been credited with being the first Turkish newspaper to have had a photograph accompany a news report.[45] Various histories of the Turkish press state that such innovations were how *Hürriyet* increased its initial circulation of 200,000 to half a million by 1965 and one million by 1969.[46] Despite the number of high-circulation newspapers and illustrated magazines printed in Istanbul throughout the 1950s, however, a countrywide distribution system was not established until the 1960s.

Hayat (*Life*) magazine was of particular importance in the development of photojournalism in Turkey.[47] This was not the only similarity it had to the famed *Life* magazine of the United States (1936–72); *Hayat* occasionally borrowed content from *Life* as well as from *Paris Match*. It immediately became a success story with a record circulation and a high "pass-along" rate, suggesting it may have entered most homes in the city. *Hayat* retained its market position by maintaining a low price, appealing content, and high advertising revenue until 1978, when it ceased publication, outliving its American model.

Güler was one of a new cadre of photojournalists whose careers were made by *Hayat*. He began at the *Yeni İstanbul* (1949–81) newspaper, joined *Hayat* in 1955, and continued working there until 1961. He gained familiarity with the work of the foremost photojournalists in the field through *Hayat*'s contacts with European and American magazines and photo agencies, briefly becoming a member of the Magnum Photo Agency in 1961. His photographs were used in *Hayat* for photo essays, as illustrations for news stories, and for advertising purposes.

Güler's oeuvre

Only some of Güler's early work has resurfaced in the present day, and these images were ones that remained largely detached from their subject matter. However, it is precisely this lack of didactic potential that makes them valuable for imagining the return of Old Istanbul in the future. A comparison of two photo essays coproduced by Güler, one from 1959 and the other from 1969, illustrates this point well. Both departed from mainstream reformist photographic documentation of the city, as described above. The first was published in three consecutive issues of *Hayat*; in it, Güler and his colleague, Orhan Tahsin, used ethnographic methods to research everyday life in the squatter settlement of Taşlıtarla. The second was published as a ten-day serial in the daily newspaper *Akşam*; it invited the public to engage in a socialist *flânerie* through some of the poorest quarters in the city. Only images from the second essay have found their way into contemporary publications that nostalgically recall Istanbul's cosmopolitan past.

Güler and the writer Orhan Tahsin conducted 45 days of participant observation in Taşlıtarla from October 1, 1959 to November 16, 1959. They roomed with an immigrant family, hiding their real identity as journalists (see Figure 2.7). The publication of the resulting twelve-page photo essay was spread over three weeks, with the first installment appearing in *Hayat*'s December 25, 1959 issue.[48]

Notably in terms of layout, the outside vertical edges of the last two installments were filled with commercial advertisements for canned food, insurance, and romance novels. This framing implies that the editors had an expected audience in mind: middle-class female homemakers. According to art historian Erica Doss, writing on *Life*, the "juxtaposition of 'instructive' articles and photo-essays" and advertisements in the layout of the magazine "relieved anxieties regarding *Life*'s educational imperative."[49] A similar functionality can be attributed to *Hayat*'s use of advertising. In the case of the Taşlıtarla essay, the reader is invited to "touch" the promise of consumerism as the piece concludes, setting oneself apart from the impoverished subjects of the essay.

The first installment likens Taşlıtarla to a rural village. Güler and Tahsin note the primitive working conditions of the settlement, with women laboring on the

Figure 2.7 Detail from the first spread of the Taşlıtarla piece. Captions to the photographs read: "*Hayat*'s journalists in Taşlıtarla: *Hayat* journalists (Ara Güler to the left and Orhan Tahsin to the right) live in a room on 45 Gülsevenler Street, sleep on a floor mattress at night, take their daily notes, and organize their films." And "Taşlıtarla: Taşlıtarla of 100,000 population extends on one side to Küçükköy, on one side to Rami, and on another side to Sağmalcılar village. Most of the houses are single story. Excluding the 2052 émigré houses [*göçmen evi*, built by the government for exiles from the Balkans], the remaining 18 thousand homes have been built without planning [and permits]."

Orhan Tahsin and Ara Güler, "Taşlıtarla'da 1.5 Ay Yaşadım," *Hayat*, no. 1, January 1, 1960, 20.

earthen ground, drying foodstuffs displayed openly, and the presence of such urban novelties as squatter homes built in two hours. In the second, they discuss the various personalities of Taşlıtarla. To their surprise, they observe how doctors, drivers, writers, and religious leaders seem to cohabit there peacefully. In the third and final installment, titled "Neighborhood with Seven Colors" (Yedi Renkli Mahalle), they introduce their hosts, Bulgarian Turks who, contrary to expectations, live a decent family life (see Figure 2.8). Güler and Tahsin note and relate their observations of everyday life in full detail, as if to correct misconceptions about the settlement.

Taşlıtarla (translates as rocky field; known today as Gaziosmanpaşa, one of Istanbul's largest and poorest district municipalities) originated as a state-sponsored resettlement community for Bulgarian Turks immigrating to Turkey (1950–1) following the implementation of assimilationist policies there. However, rural-to-urban migrants from Anatolia joined the Bulgarian Turks, and Taşlıtarla soon grew into a squatter settlement. At the time the essay was produced, the issue of migration was complicated by issues of national identity and rights to the city.

Figure 2.8 Images from the "Neighborhood with Seven Colors" installment: the photographs depict, clockwise from left, the interior of their host family with one of the journalists, Tahsin, in the middle with one of the children on his lap; the mother with one of her children; an artisan from Yugoslavia renovating old furniture; a secondhand store selling Christian iconography; tinsmithing outdoors; and children playing hula-hoop. The advertisement columns to the sides of the spread belong to romance novels to the left and insurance to the right.

Orhan Tahsin and Ara Güler, "Taşlıtarla'da 1.5 Ay Yaşadım: Yedi Renkli Mahalle," *Hayat*, no. 3, January 15, 1960, 10–11.

Migrants from the Balkans were allocated land and sometimes basic housing units, but migrants from Anatolia were totally left on their own. Newspapers and popular magazines of the day served as a medium through which the literate public discussed migration and through which it is possible to trace main lines of contention.

In their work, Güler and Tahsin present the residents in Taşlıtarla as a diverse group of interesting individual types. Güler preferred to photograph them at eye level, employing a strategy generally used to encourage viewers to identify with the subjects. The strategy also helped diminish the impact of power relations within the community and between the state and the community. The images and text in "Taşlıtarla" thus dwell on diversity and alterity. The photo essay challenges common assumptions about the migrants by portraying them as heterogeneous in terms of ethnic and linguistic origin as well as occupation.

In their celebration of diversity, Güler and Tahsin try to undo prejudices about the kinds of people who live in the settlement. This becomes especially clear when their work is compared with more mainstream depictions and discussions of migrants appearing in newspapers and other cultural outlets at the time. Part of rebutting these false impressions involved challenging the assumed homogeneity of the migrants. Thus, Güler and Tahsin reveal how one type of urban heterogeneity in Istanbul (the presence of Christian and Jewish populations of different ethnicity) is being replaced by another that is equally, if not more, diverse in its "languages." However, while celebrating the diversity of this new place, the journalists also treat it as a "foreign" land. They view the settlement of Taşlıtarla as existing outside Istanbul, which indicates that they do not see its residents as among their reading public.

One of the most interesting pictures in the essay is of a makeshift second-hand store located across from the local police headquarters. A reclining bed frame and several stacked wooden chairs are spread alongside the street, and in the foreground, a child examines a series of framed religious images. Güler and Tahsin note that these are Christian relics, but they observe there are hardly any Armenian, Greek, or Jewish residents left in the area. Hence, these objects must be there because their previous owners have hastily left the city, and their belongings have ended up in such stores.

The importance of this element of the essay assumes greater importance when one considers that the newspapers paid attention to the Muslim immigrants from the Balkans and rural migrants from Anatolia, but was surprisingly silent about the exodus of non-Muslim minorities (except for condemning the riots that took place on September 6–7, 1955). Güler and Tahsin's work thus not only sought to counteract bourgeois fears and undo prevalent notions of the poor as a homogeneous category but also presented an alternative to the increasingly popular romantic affection for a "Turkish Istanbul"—cultivated by several prominent Istanbul-based intellectuals as a critique of Ankara's hegemony and popularized most effectively in 1953 as part of the celebrations surrounding the five hundredth anniversary of the conquest of the city by the Ottoman Turks.

Ten years after the Taşlıtarla essay for *Hayat*, together with the famed author Çetin Altan, Güler carried out a three-week expedition across the city for the prominent Turkish daily *Akşam*. In contrast to the Taşlıtarla essay, this photo essay, "Here it is Istanbul!" (Al İşte İstanbul!), does not try to identify its subjects or comment on social makeup.[50] Starting with the garbage dumps by the city's ancient walls in Topkapı, Güler and Altan drove to a different squalid part of the city each day and walked around without a predetermined itinerary, noting and framing what they observed. It can be inferred that the people they encountered did not belong to their reading public because they failed to recognize Altan, who was a member of the parliament at that time and a prominent public intellectual. In fact (as seen in Figure 2.9), above the title *Akşam*, the paper advertises its own promotional lottery for a brand-new (summer, second home) apartment.[51] Thus, while the paper depicts the housing conditions of the urban poor in the serialized photo essay, it regards its readership as having access to much better housing conditions, worlds apart from the subjects of the photo essay. This juxtaposition then serves to further promote the newspaper via contrasting housing situations.

Bir kışlık bir yazlık apartman dairesi ve binlerce hediye

ÇEKİLİŞ BUGÜN

KAZANANLARIN LİSTESİ SALI GÜNÜ YAYINLANACAK.

AKŞAM

TÜRKİYENİN EN ESKİ GAZETESİ

MAYIS 25 1969 PAZAR

Azizoğlu, DP'lileri YTP'ye çağırdı

«Ordu rejime sadık olduğunu göstermiştir»

Demirel: "Biz karşı ihtilâl partisi değiliz,,

İnönü'nün başkanlığında toplanan Merkez Yönetim Kurulu tebliğinde siyasi affın sözde devrimcilerle, Demirel tarafından önlendiği açıklandı.

CHP aftan vazgeçmiyor

Başbakan ile Aydın Yalçın tartıştı

Genç kızlarımız rehberlik yapacak

Ege'de «Bizim Ev» 9 şube kuruyor

Odalar Birliği Kongresinde Hükûmet politikası yerildi

2 YAZLIK DAİRE kuponu

Komandolar Ankara'da 9 Işık Yürüyüşü yaptı

21 yaşındaki İngiliz kadınına tecavüz eden komisyoncu tutuklandı

Üzmen, kazalara karşı dinden faydalanıyor

Çetin Altan

AL İŞTE İSTANBUL!

fotoğraflar: ARA GÜLER

Yaşarken yazdığımız şehir — Penton yahut Avarlar Kapısı — Surlar ve çöplükler — Bir sütlhane — Ethem Baba — At iskeletleri...

Willy Brandt basın toplantısı yaptı

B. Almanya, Boğaz Köprüsü için 40 milyon mark kredi veriyor

GÖRÜNÜŞ

Asayiş Bakanı

İnönü'ye 21 Temmuz'a ayn inecek

Figure 2.9 The result of Çetin Altan and Ara Güler's three-week expedition across the city was published in the daily *Akşam* starting on May 25, 1969, with the title of "Al İşte İstanbul!" (Here it is Istanbul!). The text was later reprinted in paperback but without Güler's photographs. By the mid-1990s, the photographs had to be recalled from the archive for a brand new *Al İşte İstanbul!* in the form of a glossy coffee-table book, with a new sequencing of photographs each printed full page.

Çetin Altan and Ara Güler, "Al İşte İstanbul!," *Akşam*, May 25, 1969.

Altan's text and Güler's photographs sought to call more fortunate readers into action but not necessarily to reform the subjects in the photographs. Altan's text was critical of an uninformed, blasé attitude and encouraged readers to get to know Istanbul. Bringing class, rather than national identity, to the debate, he exclaimed: "The classes of the same society are living unaware of each other."[52] Drifting through the city, according to Altan, would help individuals acquire a fresh awareness of the country's "underdevelopment." This was a term that had come into local use at the end of the 1960s and was associated with neo-Marxist dependency theory, which argued that the poverty and economic stagnation characteristic of countries that were late to modernize was a result of their historical exploitation by advanced capitalist states.

The text vocalized Altan's personal desire for the reorganization of class relations by technocrats. The image captions spoke on behalf of the subjects, arguing, for instance, that "They are not interested in Istanbul, nor is Istanbul interested in them ..."[53] (caption to Güler's photo used on the cover page, seen in Figure 2.9). This obviously reflected Altan's own disinterest in the individuals portrayed in the photographs; there are no clues in the text to show if he engages them in dialogue—he merely comments on them. After the duo had undertaken a total of twenty excursions together, Altan wrote his essay about what he considered to be the pressing social problems of the city and critiqued Istanbul's urbanism. He called for the reorganization of small-scale production and its relocation from the city center to factories on its periphery. He also called for the expropriation of the shores of the Bosphorus and the construction of Ataköy-type mass-produced, large-scale housing estates for the masses. Altan thought not only that such projects could resolve social injustice and spatial fragmentation in the city but also that they could drive Turkey's modernization. The concentration of squatter settlements and the new urban poor in Istanbul served to argue, in this photo essay, that modernization had failed and that there was a need to adopt a different path for development—a path that would ensure redistribution of resources in a planned and authoritarian city.

What is interesting about this initial presentation of the work in *Akşam* is that without Altan's text, Güler's photographs would not have carried any of these meanings. Indeed, in his position as photographer of the city, Güler's photos can better be compared to those of Eugène Atget in Paris. Self-admittedly much influenced by Henri Cartier-Bresson and his "pictures on the run" (*images à la sauvatte*), Güler experimented with many different approaches to street photography over the years. Some of his pictures, such as the "wooden houses and children in the gypsy quarter" (Şişhane, 1969; no. 27 in *Lost Istanbul*), resemble the "in your face" approach of William Klein. This image of children playing on the street also brings to mind Aaron Siskind's "Most Crowded Block" from *Harlem Document, 1932–1940*. Güler's photograph of "the horse cart and tram at Sirkeci on a winter day" of 1956 (p. 53) is reminiscent of Alfred Stieglitz's "The Terminal." In addition, Güler's images of laborers recall photographs by Lewis Hine. In particular, Güler's "boy working at the machine at the repair wharf" of 1969 (p. 136) has a composition similar to Hine's 1920 "Power house mechanic working on steam pump." The subject matter of the child worker, of course, reminds one of Hine, too.[54] Güler was clearly aware of the history of the field of photography and connected to its currents.

The first picture of the ten-day series depicts a decaying inner-city neighborhood. Children inhabit the foreground of the picture (see Figure 2.10). The middle ground is occupied by several women and lines of clothing hung out to dry on an earthen street. Crumbling wooden houses are situated in the background of the photograph. Although this photograph displays poverty, its intent is not to expose the misfortune of impoverished living conditions. Poverty does not seem to weigh on the children or the few women living in this environment. One of the boys almost throws himself at the photographer; his eyes fixed on the camera, he is ready to become its subject. The viewer is thus invited to take an interest in the subjects, but not necessarily to get to know them or to take action. The photograph does not even try to persuade its audience to conclude that such conditions should change. All it does is evoke a generalized compassion. In this way, it departs from the reformist photographs of the period, which intentionally juxtaposed the

Figure 2.10 Altan and Güler's photo essay was introduced with this photograph.
Çetin Altan and Ara Güler, *Al İşte İstanbul!* (Istanbul: YKY, 2005), 150.

"traditional" with the modern, with the aim of motivating viewers to be proactive in the modernization of the city and its unmodern inhabitants.

The text Altan contributed to this joint effort to document the city in 1969 was reprinted several times during the 1970s and 1980s, but without Güler's photographs. However, in the mid-1990s, the text had become so dominant that the photographs had to be recalled from the archive for a brand-new edition of *Here it is Istanbul*, published in 1998 in the form of a glossy coffee-table book. This hardcover edition is available today in the Istanbul sections of many bookstores around the city. Its cover differs from previous paperback editions of the book printed without Güler's photos: These featured a male migrant (1970) or a father and a son in a garbage can (1980).[55] The cover of the 1998 version and later editions is more abstract, featuring a full-bleed image of handprints (perhaps on dirt, perhaps on drying mortar) printed in a monochromatic hue. This is the last image of Güler's photo essay.

In its glossy new reincarnation, the original layout of the newspaper is abandoned, the sequence of the photographs is altered, and each is printed full-page, while the lengthy text uneasily fills the space between. The text clearly does not reflect nostalgia for the city's lost cosmopolitan character. On the contrary, Altan refers to the former cosmopolitans of the city as "compradors"; and he wants the decaying Ottoman fabric, with its wooden houses, to collapse so it can be replaced with fresh mass-housing sites. Yet, while the new coffee-table edition suspends Altan's text between the photographs, this neither alters its message nor the newly invented legacy of a multicultural past presented by the photographs.

The recirculation of selected urban scenes

Güler's photographs were originally taken to provide material for an expanding print media and in relation to dramatic physical and social transformations in the city during the 1950s and 1960s. However, the reframing of Güler's work since the early 1990s, through a selection of his photographs from this earlier moment, has a much different purpose—to promote a popularized longing for the multiethnic, multi-religious past of the city.

A typical example is Güler's photograph of two boatmen, dated 1956 in various sources, which adorns the cover of his 1994 *Lost Istanbul*. To give an example of how Güler's photographs are signified, it is featured on the cover of another author's oral history book that tries to recover the memories of a multi-religious, multicultural city.[56] Neither the latter 2002 book's title, "Yesterday's Istanbul: The Dissolution of a Multi-religious, Multi-linguistic Mosaic," nor its content is about these boatmen. The photograph does not communicate in itself anything specific about the two people, their religion, or their language. The viewer can hardly make out the figures' physical features, which are backlit and discernable only as silhouettes. The figures' posture and attire only suggest their socio-economic status: working class and poor. Their loose jackets and flat caps (as opposed to rimmed hats) situate them in a loose temporality. The figures can be identified as boatmen because of their positioning in front of the two small boats, the silhouettes of which merge with the boatmen's. This dark foreground is set against a grey middle ground consisting of the Golden Horn and the Old Galata Bridge, and a background of Yeni Cami, hovering right above the figures with its domes and minarets, yet veiled by heavy fog. This is a city where small boats still operate, a city to which the electrification of streetlights may not have yet arrived. Consumed today, the photograph is nostalgic of bygone era, but it is not precise in its message. Whatever or whichever time one is nostalgic of, the photograph allows the viewer to project just that.

The appeal of Güler's photographs is thus not necessarily derived from their relation to an original subject or from the texts that may or may not accompany them, but from their potential to evoke human compassion without engagement. Güler's recycled photographs do not lend themselves to a programmatic reading. Hence, his image of a child examining Christian relics in a street in Taşlıtarla has no place in this particular album because it unveils both the "depopulated" nature of current cosmopolitanism and the heterogeneity of the emerging new social composition of the city, which is ever present today.

Photographic media and memory

Looking at old black-and-white photographic reproductions of the city's past such as Güler's, viewers are transported to a past they did not necessarily experience; or, reciprocally, this other Istanbul moves toward present-day viewers and touches them. Black-and-white images act as "melancholy objects," to borrow a phrase from Susan Sontag, who was one of the earliest critics to connect photography to nostalgia, referring to the black-and-white work of Eugène Atget and Brassaï (Gyula Halász) in documenting the disappearing "Old Paris."[57] She says:

> Cameras began duplicating the world at that moment when the human landscape started to undergo a vertiginous rate of change: while an untold number of forms of biological and social life are being destroyed in a brief span of time, a device is available to record what is disappearing. The moody, intricately textured Paris of Atget and Brassaï is mostly gone. Like the dead relatives and friends preserved in the family album, whose presence in photographs exorcises some of the anxiety and remorse prompted by their disappearance, so the photographs of neighborhoods now torn down, rural places disfigured and made barren, supply our pocket relation to the past.[58]

This connection to the family album is remarkable. Just as a child who comes to "remember" moments captured in photographs because of multiple exposures to the same images in viewing and re-viewing the family album, but may have no other lived memories of the continuum of events leading up to and following the shot, certain city photographs come to substitute for actual lived memories. These are fundamentally different than lived memories in the way they evoke mostly positive, sympathetic feelings. Sontag explains, "photographs turn the past into an object of tender regard, scrambling moral distinctions and disarming historical judgments by the generalized pathos of looking at time past."[59]

Another theorist to discuss the role of media representations in the way we look at the past was Fredric Jameson. Writing in 1984, Jameson defined "nostalgia mode" as a symptom of a crisis in postmodern thought about historical imagination.[60] In contrast to such negative interpretations of media memories, equating nostalgia with forgetting or amnesia, scholars have since pointed out the futility of drawing dualistic boundaries between real and virtual memories and have started exploring the relationship between media and memory. Marianne Hirsch coined the term "postmemory" to suggest that the perceived ideal of family photographs can be powerful for both personal and cultural memory.[61] Others have suggested media and memory are constitutive; or memories are mostly mediated and in fact, media feeds memory "fever" (hence the intensified desire to record lived experiences through technological reproduction, memorials, and museums).[62] In turn, personal or mass media provide layers of memory that the postmodern experimental self can flexibly draw from. Celia Lury identifies a shift from aesthetic culture to "prosthetic culture." She believes that a plural society, ordered by variety, is being supplanted by a post-plural society, ordered by diversity. In this new society, the self as possessive individual is being replaced by the experimental self, which requires media representations, such as photographs, for its narration.[63] Building on Celia Lury's notion of "prosthetic culture," Alison Landsberg makes an optimistic argument for "prosthetic memory": that mediated memories can be progressive, allowing individuals to potentially make counterhegemonic readings of mass media representations, to develop empathy for other peoples' conditions, which may in turn become the basis of new counterpublics.[64]

Güler's photographs of Old Istanbul have thus turned into mediated memories through circulation in publications, exhibitions, book covers, and as decorative prints in cafés and domestic spaces. In his memoir titled *Istanbul*, Nobel Laureate Orhan Pamuk includes many of Güler's Istanbul photographs (to my count, sixty-two of two hundred). The last photograph is one of Pamuk and Güler in front of a slide table, choosing images. This collaboration expands into Pamuk's house museum—where the gift shop sells only Pamuk's work with the exception of Güler's. Pamuk does not take issue with the prevalent recirculation and commodification of Güler's Istanbul photographs. Yet, he admits: "I have seen some of Güler's photographs so many times that I now confuse them with my own memories of Istanbul."[65]

Interestingly, this statement is similar to another the novelist makes in his memoir, in relationship to old black-and-white Turkish films, again from the 1950s and 1960s, that are being recycled on TV and other media.[66] Both are black-and-white, "old" media. Pamuk is cognizant of the role of black-and-whiteness, and elaborates on it in a layered manner in his *Istanbul* memoir-cum-urban history, in a chapter entitled "Black and White."[67] Here the novelist introduces the notion of seeing the city in another modality. He suggests: "(t)o see the city in black-and-white is to see it through the tarnish of history: the patina of what is old and faded and no longer matters to the rest of the world."[68] This mode of seeing metaphorically registers the city's loss of (cultural) color; that is, its former multicultural, multi-religious, and multi-linguistic complexity.[69]

Pamuk's is "autobiography as cultural criticism," a genre where the intellectual writes of his earlier life and meshes that with broader interpretations of class, culture, and history.[70] In discussing the novelist's use of photographs in the context of such autobiographies, Gabriel Koureas explains that "photography has become a key characteristic of revisionist autobiographies" to help distinguish between different selves, e.g. Orhan in the photograph, Orhan the author—pointing to the role of photography as a prosthesis for the experimental self.[71]

While it is palpable how "experimental" subjectivity works in Pamuk's memoir with the aid of photographs, it is less so at a cultural, societal level: Does Güler's work fulfill Landsberg's optimistic argument about prosthetic memories? Do these black-and-white photographs that came to represent a colorful Old Istanbul in fact make the city more diverse, or at least open to diversity? To a degree, they have.

Non-Muslim citizens of the city became, in their absence, central to Istanbul's imagined and marketed societal color. As part of the 2010 European Capital of Culture events in Istanbul, the city's multicultural past was once again on display. Interestingly, to coincide with this year of celebration, Ara Güler published a new book of fifty-six selected photographs, titled *Armenian Fishermen at Kumkapı*—this was a single volume in three languages: English, Turkish, and Armenian. The novelty was that the people in the photographs were identified as Armenian citizens in a fishing community, soon to be displaced by urban renewal. One may argue that this is staged multiculturalism, that the book still does not explore what happened to Armenians or to this particular Armenian fishing community. The photographs and the accompanying essay had originally been published in a local Armenian-language newspaper *Jamanak* in 1952 as a six-day series. In contrast, the photographer's prior books, such as *Lost Istanbul*, and exhibitions on the city had decontextualized the photographs as well as their original contexts and subjects. The publication of *Armenian Fishermen* does not yet parallel an improvement in the lives of the already-dwindled numbers of non-Muslim citizens in the city today. Yet, it points to a political opening molded very much by nostalgia. Quite unexpectedly, black-and-whiteness as a nostalgia mode opens up space, albeit a very limited space, for the memories of the city's former Armenian residents.

Güler's black-and-white photographs of Old Istanbul have been instrumental in imagining the future of the city as a socially and culturally colorful open city. The photographer's active curatorial intervention or engagement is key to the limited number of photographs that recirculate from his earlier oeuvre. There is a tension in the reception of the work; by critics and enthusiasts, the work is viewed and consumed as a means of resistance to rampant urban transformation; yet by city marketers and governments, it is viewed as a picture-perfect depiction of how much the city has improved. Through dissemination in books, exhibitions, and exhibitionary spaces, the photographs have turned into mediated memories of moments viewers may not have necessarily experienced in person. Thus, the photographs have been put to many different uses, including the narration of new selves.

The circulation of black-and-white pictures in contemporary media, be it photographs or cinema films, provide "prosthetic memories" for the imagining of Old Istanbul for the larger public. This larger public may not have personal or transmitted lived memories of that Istanbul and instead have to rely on technologically reproduced representations such as photographs, films, museums, theme parks, and other simulations to experience it. In both cinematic and photographic works, the recirculation of old black-and-white images of the city reflects an effort to affirm that Istanbul has left behind its provincial, poorer past, has moved on, and is ready for its future as a colorful European city. It is also in this sense that the Turkish state, through its various branding agencies, turns to Güler's work to promote Turkey to Europe.

Notes

1 Svetlana Boym, *The Future of Nostalgia* (New York: Basic, 2001), xiii.

2 "Ara Güler'den, Belediye'ye Sergi Tepkisi," *CNNTürk*, last modified on Mar 6, 2008, http://www.cnnturk.com/2008/kultur.sanat/diger/03/06/ara.gulerden.belediyeye.sergi.tepkisi/435105.0/index.html.

3 Christopher Phillips, "The Judgment Seat of Photography," *October* 22 (Autumn 1982): 27–63; Rosalind Kraus, "Photography's Discursive Spaces: Landscape/View," *Art Journal* 42, no. 4, The Crisis in the Discipline (Winter 1982): 311–19; Abigail Solomon-Godeau, "Canon Fodder: Authoring Eugène Atget," in *Photography at the Dock: Essays on Photographic History, Institutions, and Practices* (Minneapolis, MN: University of Minnesota Press, 1991), 28–51.

4 Douglas Crimp, *On the Museum's Ruins* (Cambridge, MA: MIT, 1993), 74.

5 The literature on this topic is vast. For an earlier and succinct account of this transformation, see this classic article: David Harvey, "From Managerialism to Entrepreneurialism: The Transformation in Urban Governance in Late Capitalism," *Geografiska Annaler. Series B, Human Geography* 71, no. 1, The Roots of Geographical Change: 1973 to the Present (1989): 3–17.

6 Cover illustrates a dossier with contributions from Abdullah Kuran, Enis Batur, and Alpay Kabacalı, and blurbs from Murat Belge, Zeynep Avcı, Hulki Altunç, Ahmet Turhan Altıner, and Mario Levi. "Kitaplara Sığınan İstanbul: Umurlarında mı İstanbul?" *Milliyet Sanat*, no. 286 (April 15, 1992), 12–19.

7 See the website of the organization for a list of founders and activities. www.tarihvakfi.org.tr/.

8 Ara Güler, *A Photographical Sketch on Lost Istanbul* (Istanbul: Dünya Yayınları, 1994). This book was published simultaneously in English and Turkish, the latter with the title *Eski İstanbul Anıları (Memories of Old Istanbul).*

9 Ministry of Foreign Affairs of the Republic of Turkey, official website: "Comprehensive Cultural Activities Recently Organized Abroad," accessed Jan 10, 2017, http://www.mfa.gov.tr/years-and-seasons-of-turkey-recently-organized-in-other-countries.en.mfa.

10 Mike MacEacheran, "Istanbul's New Museum Manifesto," *BBC Travel*, last modified December 20, 2012, http://www.bbc.com/travel/feature/20121217-istanbuls-new-museum-manifesto.

11 Sevinç Özarslan, "Talimhane'de Bir Ara Otel," in *Zaman*, Saturday Supplement, October 14, 2006, accessed January 10, 2017, http://cdncms.zaman.com.tr/2006/10/14/cumaertesi.pdf.

12 Point Hotel, "Art Inside," accessed January 11, 2017, http://www.pointhotel.com/point-hotel-taksim/art-inside.aspx.

13 Özarslan, "Talimhane'de Bir Ara Otel."

14 For example, this interview: Hacer Adıgüzel, "Ara Güler ile Sanat, Fotograf ve İstanbul'a Dair," in *İstonbul*, no. 8 (Jan.–Mar. 2003): 36–9.
15 Paul Grainge, *Monochrome Memories: Nostalgia and Style in Retro America* (Westport, CT: Praeger, 2002), xiv.
16 Pierre Nora, "Between Memory and History: Les Lieux de Mémoire," *Representations*, no. 26, Special Issue: Memory and Counter-Memory (Spring 1989): 7–24.
17 The first quote is from Siegfried Kracauer, "Photography," in *The Mass Ornament: Weimar Essays*, trans. Thomas T. Levin (Cambridge, MA: Harvard University Press, 1995), 50, quoted in Geoffrey Batchen, *Forget Me Not: Photography and Remembrance* (New York: Princeton Architectural Press, 2004), 16. The second is from Roland Barthes, *Camera Lucida: Reflections on Photography* (London: Vintage, 1982), 91, quoted in Batchen, *Forget Me Not*, 15.
18 Batchen, *Forget Me Not*, 98.
19 Craig Calhoun, "Belonging in the Cosmopolitan Imaginary," *Ethnicities* 3, no. 4 (2003): 531–68.
20 Pheng Cheah, "Introduction Part II: The Cosmopolitical—Today," in *Cosmopolitics: Thinking and Feeling Beyond the Nation*, eds. Pheng Cheah and Bruce Robbins (Minneapolis, MN: University of Minnesota Press, 1998), 21.
21 Roel Meijer, "Introduction," *Cosmopolitanism, Identity and Authenticity in the Middle East* (Richmond, UK: Curzon, 1999), 1–11, quoted in Benton Jay Komins, "Depopulated Cosmopolitanism: The Cultures of Integration, Concealment, and Evacuation in Istanbul," *Comparative Literature Studies* 39, no. 4 (2002): 360–85.
22 For example: Edhem Eldem, "Batılaşma, Modernleşme, ve Kosmopolitanism: 19. yüzyıl sonu ve 20. yüzyıl başında İstanbul," in *Osman Hamdi Bey ve Dönemi: Sempozyumu, 17–18 Aralık 1992*, ed. Zeynep Rona (Istanbul: Tarih Vakfı Yurt Yayınları, 1993), 12–26.
23 Sami Zubaida, "Cosmopolitanism and the Middle East," in *Cosmopolitanism, Identity and Authenticity in the Middle East*, ed. Roel Meijer (Richmond, UK: Curzon, 1999), 15.
24 Daniele Archibugi and Mathias Koenig-Archibugi, "Globalization, Democracy, and Cosmopolis: A Bibliographical Essay," in *Debating Cosmopolitics*, ed. Daniele Archibugi (London: Verso, 2003), 281.
25 Komins, "Depopulated Cosmopolitanism."
26 Ayşe Öncü, "Istanbulites and Others: The Cultural Cosmology of Being Middle Class in the Era of Globalism," in *Istanbul: Between the Global and the Local*, ed. Çağlar Keyder (New York: Rowman & Littlefield, 1999), 95–119.
27 Ibid., 96.
28 Stephen Kinzer, "Turkey's Passionate Interpreter to the World," *New York Times*, Arts Section, April 13, 1997, accessed January 11, 2017, http://www.nytimes.com/1997/04/13/arts/turkey-s-passionate-interpreter-to-the-world.html.
29 For an account of the relationship between cosmopolitanism and nationalism in pre-1950s republican İstanbul, see: Hakan Kaynar, "Zıt Kardeşler: Kozmopolitlik ve Milliyetçilik," in *Projesiz Modernleşme: Cumhuriyet İstanbul'undan Gündelik Fragmanlar* (İstanbul: İstanbul Araştırmaları Enstitüsü, 2012), 255–91.
30 Rıfat Bali, *Tarz-ı Hayattan Life Style'a: Yeni Seçkinler, Yeni Mekanlar, Yeni Yaşamlar* (Istanbul: İletişim Yayınları, 2002), 135–6 (author's translation).
31 Ali Müfit Gürtuna, "İstanbullu Olma Bilinci," *İstonbul* (April–June 2002): 1–2. See: Uğur Tanyeli, "Ben İstanbullu Değilim!" *Arredamento Mimarlık*, no. 161 (September 2003): 7–8.
32 Jacques Derrida, *On Cosmopolitanism and Forgiveness*, trans. Mark Dooley and Michael Hughes (London: Routledge, 2001).
33 David Lowenthal, *The Past is a Foreign Country* (Cambridge: Cambridge University Press, 1985).
34 Tahire Erman, "Becoming 'Urban' or Remaining 'Rural': The Views of Turkish Rural-to-Urban Migrants on the 'Integration' Question," *International Journal of Middle East*

Studies 30, no. 4 (1998): 541–61; Tahire Erman, "The Politics of Squatter (Gecekondu) Studies in Turkey: The Changing Representations of Rural Migrants in the Academic Discourse," *Urban Studies* 38, no. 7 (2001): 983–1002; and Tansı Şenyapılı, "Gecekondu Olgusuna Dönemsel Yaklaşımlar," in *Değişen Mekan: Mekansal Süreçlere İlişkin Tartışma ve Araştırmalara Toplu Bir Bakış, 1923–2003*, ed. Ayda Eraydın (Ankara: Dost Kitabevi Yayınları, 2006), 84–123.

35 Marshall Berman, *All That Is Solid Melts into Air: The Experience of Modernity* (London: Verso, 1983), 152–3; Gülsüm Baydar, "Silent Interruptions: Urban Encounters with Rural Turkey," in *Rethinking Modernity and National Identity in Turkey*, eds. Sibel Bozdoğan and Reşat Kasaba (Seattle: University of Washington Press, 1997), 196.

36 Yaşar Kemal, "Samsundan Istanbula Ambar Yolcuları ile Seyahat," *Cumhuriyet*, May 24, 1952.

37 Şenyapılı, "Gecekondu Olgusuna Dönemsel Yaklaşımlar," 84. Other sources that review studies of migration include İlhan Tekeli, "Yerleşme Yapıları ve Göç Araştırmaları," in *Değişen Mekan: Mekansal Süreçlere İlişkin Tartışma ve Araştırmalara Toplu Bir Bakış, 1923–2003*, ed. Ayda Eraydın (Ankara: Dost Kitabevi Yayınları, 2006), 69–83; Erman, "Becoming 'Urban' or Remaining 'Rural'"; and Erman, "The Politics of Squatter (*Gecekondu*) Studies in Turkey."

38 "Cumhuriyetin Anketi: Belediye Her Ay 500 Aç, Perisan Anadoluluya Dönüş Bileti Alıyor," *Cumhuriyet*, August 1, 1951.

39 This was a commentary more than a survey. It was published on the front page between July 29, 1951 and August 1, 1951. "Cumhuriyetin Anketi: Anadoludan İstanbula Akın Edenlerin Acıklı Akıbetleri," *Cumhuriyet*, July 29, 1951; "Cumhuriyetin Anketi: Şöförlüğü Tarlaya Tercih Edenler," *Cumhuriyet*, July 30, 1951; "Cumhuriyetin Anketi: İstanbula Yapılan Akının Öncüleri Lokantacılar," *Cumhuriyet*, July 31, 1951; and "Cumhuriyetin Anketi: Belediye Her Ay 500 Aç, Perişan Anadoluluya Dönüş Bileti Alıyor," *Cumhuriyet*, August 1, 1951.

40 Tekeli, "Yerleşme Yapıları ve Göç Araştırmaları," 78.

41 Sibel Bozdoğan, *Modernism and Nation Building: Turkish Architectural Culture in the Early Republic* (Seattle: University of Washington Press, 2001), 133.

42 Author's translation of the following caption: "Burnunun Dibinde: Şehir nizamını, vatandaş haklarını ve sağlığını, cemiyet problemlerini ilgilendiren kararların, yasakların pek çoğu yukarıda resmi görülen Belediye Sarayı içerisindeki idareciler tarafından yürütülür. Resimde görüldüğü gibi, sahipleri tarafından götürülen inekler Belediye'nin önünden ağır ağır geçiyorlar." *Yeni İstanbul*, August 2, 1962.

43 *Cumhuriyet*, March 5, 1960.

44 Ahmet Oktay, *Toplumsal Değişme ve Basın: 1960–1986 Türk Basını Üzerine Bir Çalışma* (Turkey: Bilim/Felsefe/Sanat Yayınları, 1987); Nurhan Kavaklı, *Bir Gazetenin Tarihi: Akşam* (Istanbul: Yapı Kredi Yayıncılık, 2003); and Aysun Köktener, *Bir Gazetenin Tarihi: Cumhuriyet* (Istanbul: Yapı Kredi Yayıncılık, 2004).

45 Seyit Ali Ak, *Erken Cumhuriyet Dönemi Türk Fotoğrafı (1923–1960)* (Istanbul: Remzi Kitabevi 2001), 181.

46 Ibid.

47 Engin Özdeş, *Photography in Turkey* (Istanbul: Pamukbank and History Foundation, 1999), 26.

48 The first installment appeared on *Hayat* 2, no. 168 (4th year) on December 25, 1959. The following issues appeared on January 1, 1960, and January 8, 1960.

49 Erika Doss, "Introduction: Looking at Life: Rethinking America's Favorite Magazine, 1936–1972," in *Looking at Life Magazine*, ed. Erika Doss (Washington, DC: Smithsonian Institution Press, 2001), 8.

50 Çetin Altan and Ara Güler, "Al İşte İstanbul!," *Akşam*, May 26, 1969. Photos and text reproduced in Çetin Altan and Ara Güler, *Al İşte İstanbul!* (Istanbul: YKY, 2005 [1998]).

51 From the 1950s until the 1970s, it was mainly banks, which offered apartments as prizes in lotteries to promote their business. New customers who opened accounts were entered into a draw (*çekiliş*). The practice of advertising by lottery prize apartments was picked up by other types of business, such as newspapers.
52 Altan, *Al İşte İstanbul!*, 169.
53 Altan and Güler, "Al İşte İstanbul!," 1.
54 I thank historians of photography Douglas Nickel and Mazie Harris for pointing out to me such references in Güler's work.
55 Çetin Altan, *Bir Uçtan Bir Uca ve Al İşte İstanbul!* (Istanbul: Kitapçılık Ticaret Ltd. Şirketi, 1970); and Çetin Altan, *Al İşte İstanbul!* (Istanbul: Yazko, 1980).
56 İlhan Eksen, *Dünkü İstanbul: Çok Dinli, Çok Dilli Mozaiğin Dağılışı* (Istanbul: Sel Yayıncılık, 2002).
57 Emphasis mine. Susan Sontag, *On Photography* (New York: Picador, 1977), 15–16.
58 Ibid., 16.
59 Ibid., 71
60 Fredric Jameson, "Postmodernism, or the Cultural Logic of Late Capitalism," *New Left Review*, no. 146 (July–August 1984): 53–92.
61 Marianne Hirsch, *Family Frames: Photography, Narrative, and Postmemory* (Cambridge, MA: Harvard University Press, 1997).
62 Andreas Huyssen, *Twilight Memories: Marking Time in a Culture of Amnesia* (New York: Routledge, 1995); José van Dijck, *Mediated Memories in the Digital Age* (Stanford, CA: Stanford University Press, 2007).
63 Celia Lury, *Prosthetic Culture: Photography, Memory, and Identity* (New York: Routledge, 1998).
64 Alison Landsberg, *Prosthetic Memory: The Transformation of American Remembrance in the Age of Mass Culture* (New York: Columbia University Press, 2004).
65 Orhan Pamuk, Foreword to *Ara Güler's Istanbul* (London: Thames & Hudson, 2009).
66 Orhan Pamuk, *Istanbul: Memories and the City*, trans. Maureen Freely (New York: Alfred A. Knopf, 2005).
67 Pamuk, *Istanbul.*
68 Ibid., 38.
69 Ibid., 39.
70 Nancy Miller, "Getting Personal: Autobiography as Cultural Criticism," in *Getting Personal: Feminist Occasions and Other Autobiographical Acts* (New York: Routledge, 1992), 1–30; Linda Haverty Rugg, *Picturing Ourselves: Photography & Autobiography* (Chicago, IL: University of Chicago Press, 1997).
71 Gabriel Koureas, "Orhan Pamuk's Melancholic Narrative and Fragmented Photographic Framing: *Istanbul, Memories of a City* (2005)," in *The Photo Book, from Talbot to Ruscha and Beyond*, eds. Patrizia Di Bello, Colette E Wilson, and Shamoon Zamir (London: I.B. Tauris, 2012), 211–32.

Bibliography

Adıgüzel, Hacer. "Ara Güler ile Sanat, Fotograf ve İstanbul'a Dair." In *İstonbul*, no. 8 (Jan.–Mar. 2003): 36–9.

Ak, Seyit Ali. *Erken Cumhuriyet Dönemi Türk Fotoğrafı (1923–1960)*. Istanbul: Remzi Kitabevi, 2001.

Altan, Çetin. *Bir Uçtan Bir Uca ve Al İşte İstanbul!*. Istanbul: Kitapçılık Ticaret Ltd. Şirketi, 1970.

—. *Al İşte İstanbul!*. Istanbul: Yazko, 1980.

—. *Al İşte İstanbul!*. Istanbul: YKY, 2005 [1998]).

"Ara Güler'den, Belediye'ye Sergi Tepkisi." *CNNTürk*. Last modified on March 6, 2008. http://www.cnnturk.com/2008/kultur.sanat/diger/03/06/ara.gulerden.belediyeye.sergi.tepkisi/435105.0/index.html.

Archibugi, Daniele, ed. *Debating Cosmopolitics*. London: Verso, 2003.
Archibugi, Daniele, and Mathias Koenig-Archibugi. "Globalization, Democracy, and Cosmopolis: A Bibliographical Essay." In *Debating Cosmopolitics*, edited by Daniele Archibugi, 273–292. London: Verso, 2003.
Bali, Rıfat. *Tarz-ı Hayattan Life Style'a: Yeni Seçkinler, Yeni Mekanlar, Yeni Yaşamlar*. Istanbul: İletişim Yayınları, 2002.
Barthes, Roland. *Camera Lucida: Reflections on Photography*. London: Vintage, 1982.
Batchen, Geoffrey. *Forget Me Not: Photography and Remembrance*. New York: Princeton Architectural Press, 2004.
Baydar, Gülsüm. "Silent Interruptions: Urban Encounters with Rural Turkey." In *Rethinking Modernity and National Identity in Turkey*, edited by Sibel Bozdoğan and Reşat Kasaba, 192–210. Seattle: University of Washington Press, 1997.
Berman, Marshall. *All That Is Solid Melts into Air: The Experience of Modernity*. London: Verso, 1983.
Bilsel, Cânâ, and Pierre Pinon, eds. *İmparatorluk Başkentinden Cumhuriyet'in Modern Kentine: Henri Prost'un İstanbul Planlaması (1936–1951) = From the Imperial Capital to the Republican Modern City: Henri Prost's Planning of Istanbul (1936–1951)*. Istanbul: İstanbul Araştırmaları Enstitüsü, 2010.
Boym, Svetlana. *The Future of Nostalgia*. New York: Basic, 2001.
Bozdoğan, Sibel. *Modernism and Nation Building: Turkish Architectural Culture in the Early Republic*. Seattle: University of Washington Press, 2001.
Calhoun, Craig. "Belonging in the Cosmopolitan Imaginary." *Ethnicities* 3, no. 4 (2003): 531–68.
Cheah, Pheng. "Introduction Part II: The Cosmopolitical—Today." In *Cosmopolitics: Thinking and Feeling Beyond the Nation*, edited by Pheng Cheah and Bruce Robbins, 20–41. Minneapolis, MN: University of Minnesota Press, 1998.
Crimp, Douglas. *On the Museum's Ruins*. Cambridge, MA: MIT, 1993.
"Cumhuriyetin Anketi: Anadoludan İstanbula Akın Edenlerin Acıklı Akıbetleri." *Cumhuriyet*, July 29, 1951.
"Cumhuriyetin Anketi: Belediye Her Ay 500 Aç, Perişan Anadoluluya Dönüş Bileti Alıyor." *Cumhuriyet*, August 1, 1951.
"Cumhuriyetin Anketi: İstanbula Yapılan Akının Öncüleri Lokantacılar." *Cumhuriyet*, July 31, 1951.
"Cumhuriyetin Anketi: Şöförlüğü Tarlaya Tercih Edenler." *Cumhuriyet*, July 30, 1951.
Derrida, Jacques. *On Cosmopolitanism and Forgiveness*. Translated by Mark Dooley and Michael Hughes. London: Routledge, 2001.
Doss, Erica, ed. *Looking at Life Magazine*. Washington, DC: Smithsonian Institution Press, 2001.
Eksen, İlhan. *Dünkü İstanbul: Çok Dinli, Çok Dilli Mozaiğin Dağılışı*. Istanbul: Sel Yayıncılık, 2002.
Eldem, Edhem. "Batılaşma, Modernleşme, ve Kosmopolitanism: 19. yüzyıl sonu ve 20. yüzyıl başında İstanbul." In *Osman Hamdi Bey ve Dönemi: Sempozyumu, 17–18 Aralık 1992*, edited by Zeynep Rona, 12–26. Istanbul: Tarih Vakfı Yurt Yayınları, 1993.
Erman, Tahire. "Becoming 'Urban' or Remaining 'Rural': The Views of Turkish Rural-to-Urban Migrants on the 'Integration' Question." *International Journal of Middle East Studies* 30, no. 4 (1998): 541–61.
—. "The Politics of Squatter (Gecekondu) Studies in Turkey: The Changing Representations of Rural Migrants in the Academic Discourse." *Urban Studies* 38, no. 7 (2001): 983–1002.

Grainge, Paul. *Monochrome Memories: Nostalgia and Style in Retro America*. Westport, CT: Praeger, 2002.
Gül, Murat. *The Emergence of Modern Istanbul: Transformation and Modernisation of a City*. London; New York: Tauris Academic Studies, 2009.
Gül, Mehmet Murat and Richard Lamb, "Urban Planning in Istanbul in the Early Republican Period: Henri Prost's Role in Tensions among Beautification, Modernization and Peasantist Ideology." *Architectural Theory Review* 9, no. 1 (2004): 59–79.
Güler, Ara. *A Photographical Sketch on Lost Istanbul*. Istanbul: Dünya Yayınları, 1994.
—. *Eski İstanbul Anıları*. Istanbul: Dünya Yayınları, 1994.
Gürtuna, Ali Müfit. "İstanbullu Olma Bilinci." *İstonbul* (April–June 2002): 1–2.
Harvey, David. "From Managerialism to Entrepreneurialism: The Transformation in Urban Governance in Late Capitalism." *Geografiska Annaler. Series B, Human Geography* 71, no. 1, The Roots of Geographical Change: 1973 to the Present (1989): 3–17.
Haverty Rugg, Linda. *Picturing Ourselves: Photography & Autobiography*. Chicago, IL: University of Chicago Press, 1997.
Hirsch, Marianne. *Family Frames: Photography, Narrative, and Postmemory*. Cambridge, MA: Harvard University Press, 1997.
Huyssen, Andreas. *Twilight Memories: Marking Time in a Culture of Amnesia*. New York: Routledge, 1995.
Jameson, Fredric. "Postmodernism, or the Cultural Logic of Late Capitalism." *New Left Review*, no. 146 (July–August 1984): 53–92.
Kavaklı, Nurhan. *Bir Gazetenin Tarihi: Akşam*. Istanbul: Yapı Kredi Yayıncılık, 2003.
Kaynar, Hakan. *Projesiz Modernleşme: Cumhuriyet İstanbul'undan Gündelik Fragmanlar*. Istanbul: İstanbul Araştırmaları Enstitüsü, 2012.
Kemal, Yaşar. "Samsundan İstanbula Ambar Yolcuları ile Seyahat." *Cumhuriyet*, May 24, 1952.
Kinzer, Stephen. "Turkey's Passionate Interpreter to the World." *New York Times*, Arts Section. April 13, 1997. Accessed January 11, 2017. http://www.nytimes.com/1997/04/13/arts/turkey-s-passionate-interpreter-to-the-world.html.
"Kitaplara Sığınan İstanbul: Umurlarında mı İstanbul?" *Milliyet Sanat*, no. 286 (April 15, 1992): 12–19.
Köktener, Aysun. *Bir Gazetenin Tarihi: Cumhuriyet*. Istanbul: Yapı Kredi Yayıncılık, 2004.
Komins, Benton Jay. "Depopulated Cosmopolitanism: The Cultures of Integration, Concealment, and Evacuation in Istanbul." *Comparative Literature Studies* 39, no. 4 (2002): 360–85.
Koureas, Gabriel. "Orhan Pamuk's Melancholic Narrative and Fragmented Photographic Framing: *Istanbul, Memories of a City* (2005)." In *The Photo Book, from Talbot to Ruscha and Beyond*, edited by Patrizia Di Bello, Colette E Wilson, and Shamoon Zamir, 211–32. London: I.B. Tauris, 2012.
Kracauer, Siegfried. "Photography." In *The Mass Ornament: Weimar Essays*, translated by Thomas T. Levin, 47–63. Cambridge, MA: Harvard University Press, 1995.
Kraus, Rosalind. "Photography's Discursive Spaces: Landscape/View." *Art Journal* 42, no. 4, The Crisis in the Discipline (Winter 1982): 311–19.
Landsberg, Alison. *Prosthetic Memory: The Transformation of American Remembrance in the Age of Mass Culture*. New York: Columbia University Press, 2004.
Lowenthal, David. *The Past is a Foreign Country*. Cambridge: Cambridge University Press, 1985.
Lury, Celia. *Prosthetic Culture: Photography, Memory, and Identity*. New York: Routledge, 1998.

MacEacheran, Mike. "Istanbul's New Museum Manifesto." *BBC Travel*. Last modified December 20, 2012. http://www.bbc.com/travel/feature/20121217-istanbuls-new-museum-manifesto.

Meijer, Roel, ed. *Cosmopolitanism, Identity and Authenticity in the Middle East*. Richmond, UK: Curzon, 1999.

Miller, Nancy. *Getting Personal: Feminist Occasions and Other Autobiographical Acts*. New York: Routledge, 1992.

Ministry of Foreign Affairs of the Republic of Turkey, official website. "Comprehensive Cultural Activities Recently Organized Abroad." Accessed Jan 10, 2017. http://www.mfa.gov.tr/years-and-seasons-of-turkey-recently-organized-in-other-countries.en.mfa.

Nora, Pierre. "Between Memory and History: Les Lieux de Mémoire." *Representations*, no. 26, Special Issue: Memory and Counter-Memory (Spring 1989): 7–24.

Oktay, Ahmet. *Toplumsal Değişme ve Basın: 1960–1986. Türk Basını Üzerine Bir Çalışma*. Turkey: Bilim/Felsefe/Sanat Yayınları, 1987.

Öncü, Ayşe. "Istanbulites and Others: The Cultural Cosmology of Being Middle Class in the Era of Globalism." In *Istanbul: Between the Global and the Local*, edited by Çağlar Keyder, 95–119. New York: Rowman & Littlefield, 1999.

Özarslan, Sevinç. "Talimhane'de Bir Ara Otel." In *Zaman*, Saturday Supplement. October 14, 2006. Accessed January 10, 2017. http://cdncms.zaman.com.tr/2006/10/14/cumae-rtesi.pdf.

Özdeş, Engin. *Photography in Turkey*. Istanbul: Pamukbank and History Foundation, 1999.

Pamuk, Orhan. *Istanbul: Memories and the City*. Translated by Maureen Freely. New York: Alfred A. Knopf, 2005.

—. Foreword to *Ara Güler's Istanbul*. London: Thames & Hudson, 2009.

Phillips, Christopher. "The Judgment Seat of Photography." *October* 22 (Autumn 1982): 27–63.

Point Hotel. "Art Inside." Accessed January 11, 2017. http://www.pointhotel.com/point-hotel-taksim/art-inside.aspx.

Solomon-Godeau, Abigail. *Photography at the Dock: Essays on Photographic History, Institutions, and Practices*. Minneapolis, MN: University of Minnesota Press, 1991.

Sontag, Susan. *On Photography*. New York: Picador, 1977.

Şenyapılı, Tansı. "Gecekondu Olgusuna Dönemsel Yaklaşımlar." In *Değişen Mekan: Mekansal Süreçlere İlişkin Tartışma ve Araştırmalara Toplu Bir Bakış, 1923–2003*, edited by Ayda Eraydın, 84–123. Ankara: Dost Kitabevi Yayınları, 2006.

Tanyeli, Uğur. "Ben İstanbullu Değilim!" *Arredamento Mimarlık*, no. 161 (September 2003): 7–8.

Tekeli, İlhan. "Yerleşme Yapıları ve Göç Araştırmaları." In *Değişen Mekan: Mekansal Süreçlere İlişkin Tartışma ve Araştırmalara Toplu Bir Bakış, 1923–2003*, edited by Ayda Eraydın, 69–83. Ankara: Dost Kitabevi Yayınları, 2006.

van Dijck, José. *Mediated Memories in the Digital Age*. Stanford, CA: Stanford University Press, 2007.

Zubaida, Sami. "Cosmopolitanism and the Middle East." In *Cosmopolitanism, Identity and Authenticity in the Middle East*, edited by Roel Meijer, 15–34. Richmond, UK: Curzon, 1999.

3 Cinematic memories

Old Turkish films, most shot on location in Istanbul, are viewed today across generations with tender regard because they show a city in black-and-white (monochrome), a city that was much poorer and more provincial—a city that is no more. While based on real-life events, especially news stories, they portray the city through a black-and-white vision of housing: light-filled, modern, hygienic apartments of the middle classes versus the crowded inner-city slums and squatter settlements of recent rural-to-urban migrants and working classes. The contrast between the two realms effectively communicates class conflict and it is usually resolved in favor of the latter—revealing the urban poor to be the imagined audience of many of these films. How do the films fictionalize urban experience? And what is the appeal of viewing old films in the present, when the city has transformed beyond recognition? First, I consider the return of old films and their role in the urban imaginary. Second, I situate Turkish cinema in the city as an industry broader than, but encompassing, films. Third, I explore the appeal of the medium and black-and-white as a modality of seeing. Then, I examine three feature films from around the beginning of the 1960s that are well known and continue to circulate today. These three films comment on the city's transformation via their characters' conditions and dilemmas of housing. Next, I dwell on one of the films to show how it fictionalized real phenomena in the city, namely new housing developments, infrastructural modernization, and real-estate-speculation-driven corruption. Finally, I return to and consider the appeal of these films in the present as memory objects. My discussion in this chapter complements and builds on the discussion of the parallel recirculation of select black-and-white photographs from Ara Güler's oeuvre, the focus of Chapter 2.

The return of old films and the urban imaginary

Our imaginations of cities are mediated via technologies of transportation, communication, and reproduction.[1] Among them, scholars have so far concerned themselves mostly with cinema and its privileged relationship to the modern city. Writings on cinema and the city have consistently argued that our experience of cities is increasingly mediated by, rather than merely grounded in, material space and social practice. However, the relationship between mass media, memory, and the city in this literature has not garnered due attention. Especially in fast-developing cities like Istanbul, where the built environment is rapidly changing and is complemented by population change and cultural transformation, media representations can supplant as well as transplant missing inter-generationally, handed-down, or personally experienced memories of place. Studying the relationship between media, memory, and the city necessitates focusing not only on films as display objects but also on the contexts and practices of production, circulation, and viewing.

Most studies on the cinematic city are based on two fundamental assumptions: that the modern city is the precondition of cinema's existence; and that the modern city is shaped by the cinematic imagination (for instance, the modern city is perceived through the windshield of the car, the train, or the airplane, in a state of distraction, as a montage of arrivals and departures).[2] This mode of analysis is particularly inspired by Jean Baudrillard's theorization of the simulacra and his suggestion that "the American city seems to have stepped right out of the movies."[3] Indeed, not only American cities but all cities are experienced to a great extent through their mediated images in film. This view harkens back to the writings of Georg Simmel and Walter Benjamin, who analyzed the modern city in the context of prewar Germany, and Berlin as its rapidly urbanizing industrial center, in terms of its perceptual and psychological effects. Simmel believed that people learn to negotiate and cope with changing social relations and anonymity through consumption.[4] Following Simmel's argument, Benjamin observed that the modern city, with its new structures and speeds, had brought a change in human perception; cinema's sensorial adjustment could bring about political empowerment.[5] Following this line of reasoning, much of the literature on the cinematic city has attempted to bring to light the "counter discourse" offered by films, yet with a tendency to reduce cinema to films and analyze films as texts.[6] Reflecting on the return and recirculation of old films as memory objects as well as situating cinema in the city can greatly inform the analysis of individual films, since they are inevitably viewed with the concerns of the present.

Since the late 1990s, it is possible to view the black-and-white films from the heyday of Turkish cinema in the comfort of one's home—that is, roughly from the mid-1950s to mid-1970s, dubbed "Yeşilçam" after the street on which film production companies were located in Istanbul. In 1994, a new code (Law #3984 on the Establishment of Radio and Television Enterprises and Their Broadcasts) ended the monopoly of state radio television in Turkey and helped a private sector of media channels to flourish. One of the aftereffects—switching on television and surfing the numerous new private channels—meant one could come across many of these old Yeşilçam films. One subscription channel, SinemaTURK, was devoted entirely to showing (mostly old) Turkish films, one after another. It is even possible to encounter the same film on multiple channels at once at different points in the story. In addition, many of the well-known old films have now been rereleased on DVD, and they are frequently the subjects of retrospectives in Turkish film festivals abroad. Internet streaming sites also contribute to their spread. Through TV series and DVDs, as quotations in contemporary films and art works, and as inspiration for wildly popular TV series and other forms of visual culture, black-and-white films constitute a popular archive.[7] This archive doubles as an archive of Istanbul, as the city appears not only as a location but a central character.[8]

To be specific in my reading of the "memory work" of Turkish cinema vis-à-vis Istanbul, I focus on several celebrated films from the early 1960s in relation to the local history of the medium and the modernization of the city: Halit Refiğ's *Birds of Exile* (*Gurbet Kuşları*, 1964), Metin Erksan's *Bitter Life* (*Acı Hayat*, 1963), and Ertem Göreç and Vedat Türkali's *Bus Passengers* (*Otobüs Yolcuları*, 1960).[9] These are some of the most famous films from the heyday of Turkish cinema that have received due attention within Turkish film studies and have been dubbed examples of social realism or "people's cinema." They portray experiences of the years in which they were made (i.e. they do not project into the past or the future as in period films).

The anxieties introduced in Chapter 2—in the discussion of the journalistic visual and, in particular, photographic presentations of rural-to-urban migrants and the poor in the city—are very much present in these films. As much as so, the films present the (vistas of the) city as a source of inspiration, love, and empowerment for their characters, possibly reflecting their directors' passion for the city. However, it is ultimately the question of housing that ties them together for my discussion here. These three films collectively portray and comment on social segregation in housing with an emphasis on "apartmentization" (*apartmanlaşma*). They criticize contemporaneous housing conditions fifty years ago and provide an interesting historical context for the renewed popularity of wooden historic homes, beginning in the 1970s (discussed in Chapter 4), as well as the current situation under neo-liberal policies.

Looking at old films today can also contribute to our understanding of the city as it is experienced. Shot on location and with plots loosely based on real-life stories derived from news reports, these select films are exemplary of a larger body of work that portrays the experience of Istanbul through their characters' conditions of and aspirations for housing. They present a rapidly changing city with increasing segregation based on social class, and they evoke a struggle to maintain individual innocence in the face of change. As is typical of Turkish films from the era, they tend to straddle a fine line between selling images of Istanbul and participating in a civilizing process, simultaneously training their characters in urban behavior and showing how they are able to resist the temptations of the city.

Their makers almost always side with the poor against more affluent social groups, associating the latter with moral degradation. Within the world of such films, a makeover or a winning lottery ticket can help a protagonist switch roles overnight. Eventually, the key conflict between the rural/urban or poor/rich is "resolved through the realm of fantasy."[10] However, the three films discussed here offer other kinds of resolutions, too—from promoting a communitarian model of citizenship and sending migrants to Istanbul back to their homes, to proposing a cultured enjoyment of the city's architecture and landscape as a valid alternative to participating in the consumer economy—reflecting the unique political orientations of their auteur directors. In addition to being some of the most quoted films, I discuss them together to map the imagined housing landscape of the city. However, before the individual films, cinema needs to be located in the city. The memory of cinema, as industry, is also complex and informs the way individual films are circulated and received today.

Locating cinema in the city

Cinema has played a special role in Istanbul, helping residents and audiences elsewhere in Turkey and abroad form an image of the city. If a "cinematic map" of Istanbul were to be drawn, Beyoğlu/Pera would occupy its center. The district hosted the city's first film screenings (1896) and its first (1908), as well as most prestigious, movie theaters. These were located along its main artery, the Grand Rue de Pera (İstiklal or Independence Avenue). Most film businesses during the heyday of Turkish cinema were also located in this area. Based in this district, the Turkish film industry was a multicultural enterprise. Indeed, many professionals involved in the production, distribution, and exhibition of films were of Armenian, Jewish, and Greek origin. If they were actors, they would simply use pseudonyms on screen. Second to İstiklal Avenue as a cinematic destination was Yeşilçam Street. It was on and around Yeşilçam Street that artist agencies, advertising, production, and distribution companies gathered. It was here that famous movie theaters such as the Emek, İpek, and Rüya were located. "Yeşilçam" thus came to refer not only to a place but also to a time when hundreds of films were hurriedly produced every year and, thereby, to a mode of production and the cumulative culture of local film production, distribution, exhibition, and consumption.

Domestic cinema was a late bloomer in the cultural and economic life of the city. Production expanded from a few films per year before World War II to several hundred in the 1960s, before scaling back to the dozens by the 1990s. In the immediate postwar period, as America replaced Europe as the paradigm of modernity, Hollywood films promoted American lifestyles and a star culture. By the 1960s, however, domestic products and a nascent domestic star culture were able to rival that of Hollywood. This even led the popular magazine *Ses* to provide a star map, "Local Beverly Hills," showing which stars lived in flats in the Teşvikiye-Nişantaşı area (see Figure 3.1). Using such maps, interviews, and serials, illustrated magazines and magazine supplements to daily newspapers were able to guide their readers through the homes and itineraries of celebrities, enabling the creation of an imagined shared space and time. Such coverage helped create an "imagined community" of Istanbulites.

Cinema also acted as a "public sphere" where different social groups could come together, despite increasing residential segregation.[11] A cartoon published in *Akbaba* illustrates the city's life, as miniaturized in the movie theater (see Figure 3.2). Here the city's life comes together in a movie theater replete with fragments of urban society such as a university classroom, an elite social club, a traditional puppet-show theater, a women's-only "harem," a fire in one of the balconies, a street vendor, and a rendezvous. There is even a film shoot. Clearly,

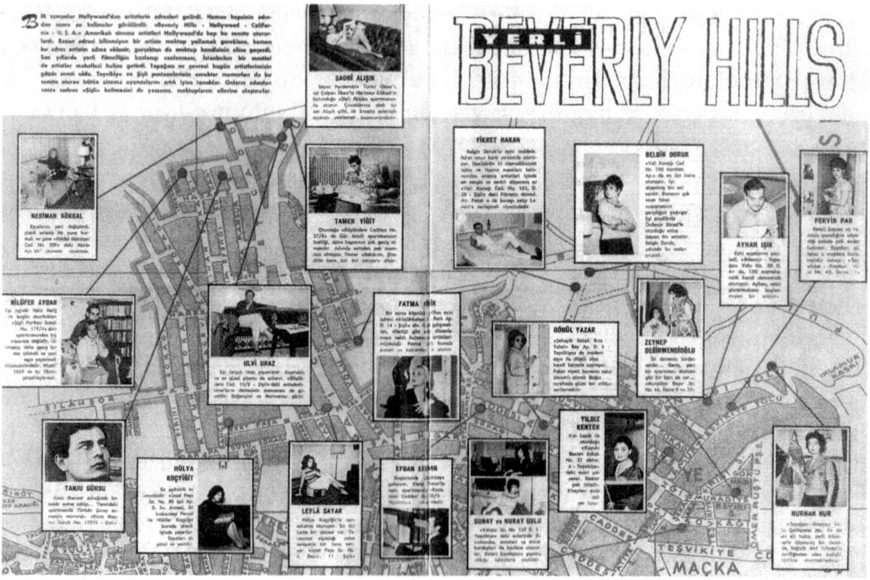

Figure 3.1 A star map of Istanbul entitled "Local Beverly Hills" furthers the feeling of a shared space and time.

"Yerli Beverly Hills," *Ses*, October 10, 1964.

Figure 3.2 Movie theater becomes a synecdoche for the city.

Cafer Zorlu and Nehar Tüblek, "Sinema Dünyası," *İstanbul* (July 1999): 110–11 (reproduced from *Akbaba*, no. 392, 1959.

this cartoon should not simply be read as a large movie theater; it is a synecdoche for the city. On the lower left are the filmmakers and critics, such as Suat Yalaz, Nejat Saydan, Nijat Özön, and Ali Gevgili, gathered around a shoot. This fragment alludes to the attraction and prevalence of outdoor film shoots in Istanbul. There is an interesting dialogue shown between several characters in the corner that reflects the ambivalence of Turkish intellectuals toward this commercially driven form of popular cinema production. Thus, one character says, "*Bir de yerli filmleri beğenmezler yahu* (See! And they do not like domestic films)," suggesting that the domestic film being shot is not that different from (and perhaps even better than) that on the screen—which is possibly a Clark Gable–Hollywood film. Another replies, "I prefer art films" (as opposed to domestic or American films).[12]

During the Yeşilçam years, production consisted largely of melodramas and comedies for a predominantly family audience.[13] Films were characteristically hastily made on shoestring budgets. They were technically low quality; most were dubbed in the studio because on-location sound recording was difficult. However, a major boon for their popularity was that they were in the Turkish language and hence accessible for domestic audiences. In addition, the audience information provided by the regional agency system used for distribution across Turkey helped customize the films and increase their numbers.[14] The medium became the major form of public entertainment, reaching mass audiences across Turkey, with half of the national audience concentrated in Istanbul.[15] Box-office figures as well as the number of outdoor screening spaces and indoor theatres gradually picked up until the mid-1970s, when a series of factors, including the nationwide spread of TV, undermined its influence.[16]

Starting in the second half of the 1990s, but especially in the 2000s, domestic production got back on its feet. The deregulation of state-controlled TV and radio, partly driven by the neoliberal agenda of privatization and partly encouraged by the EU's pressure to reform and democratize, allowed an increase in private TV channels and created work for directors who could use their earnings to fund their own films.[17] By 2006, domestic films accounted for more than fifty percent of box-office receipts. European funding schemes such as Eurimages also encouraged multinational co-productions, helped to improve production values, and supported distribution, thus rendering Turkish films more visible on an international stage.[18] Despite Istanbul's "fading out" in Turkish films produced since the mid-1990s—dubbed as New Turkish Cinema—a new generation of Eurimages-funded, transnationally produced films has taken renewed interest in the city, from Ferzan Özpetek's *Hamam* or *Steam: The Turkish Bath* (1997) to Fatih Akın's *Head-on* (2004) and *Crossing the Bridge: The Sound of Istanbul* (2005)—the latter incorporating a fragment from an old Turkish film, *Nights of Istanbul* (*İstanbul Geceleri*, Mehmet Muhtar, 1950).

Yeşilçam and its audience appeal may have been defunct in the late 1970s and '80s, but it was by no means dead. The 2000s witnessed the return of Yeşilçam films and a matching proliferation of Yeşilçam-inspired TV dramas on private channels. Yeşilçam-inspired blockbusters returned to cinemas, and revalorizing studies on Yeşilçam films grew out of communication faculties at the new private universities that began to open in Turkey in the 1990s.[19] In a country increasingly strained by neoliberal economic restructuring and troubled by the rise of sectarian identities, this popular and scholarly interest arose from a reevaluation of the way identity issues had once been dealt with in Yeşilçam films. Within local film studies, Yeşilçam films have been interpreted as "narratives of resistance" as well as "our imaginary homeland."[20]

Black-and-white visions of the city

Watching old films today, viewers are transported to a past they did not necessarily experience, or, reciprocally, this other Istanbul moves toward present-day viewers and touches them. Giuliana Bruno persuasively argues that motion pictures move through the inner space of their viewers, as well as through time, space, and narrative development—that cinematic motion involves a "haptic affective transport."[21] Their black-and-whiteness further contribute to this effect. As an aesthetic sign, monochrome modality enhances their affect.

The relationship between memory and (Turkish) film has attracted scholarly interest, mainly in relationship to New Turkish Cinema, but memory is discussed insofar as the films themselves deal with a traumatic past.[22] The recirculation of Yeşilçam films (like the old photographs discussed in Chapter 2) instead provides what Alison Landsberg calls "prosthetic memory."[23] The images of Istanbul presented in these films substitute for lived memories, especially for individuals who did not experience the city during those years because they had not been born or because they arrived in Istanbul later on. To see a city in black-and-white points, first of all, to seeing it through media, through its technological reproductions. As such, the films act as prosthesis to the workings of individual and collective memories of the city.

Perhaps the most evocative description of the memory work that old Turkish films do in the present comes from the novelist Orhan Pamuk, who writes in his memoir how they act as memory objects and provoke nostalgia:

> In the 1950s and 1960s, like everyone, I loved watching the "film crews" all over the city—the minibuses with the logos of film companies on their sides; two huge generator-powered lights; the prompters, who preferred to be known as *souffleurs* and who had to shout mightily over the generator's roar at those moments when the heavily made-up actresses and romantic male leads forgot their lines; the workers who jostled the children and curious on-lookers off the set. Forty years on, the Turkish film industry is no longer . . . they still show those old black-and-white films on television, and when I see the streets, the old gardens, the Bosphorus views, and the broken-down mansions and apartments in black-and-white, I sometimes forget I am watching a film; stupefied by melancholy, I sometimes feel as if I am watching my own past.[24]

These comments, which come from a chapter entitled "Black and White" in Pamuk's memoir, are similar to others Pamuk makes in relation to Ara Güler's old black-and-white photographs (see Chapter 2).[25] While the time frame may not be that "old," black-and-whiteness helps situate these (still or motion) pictures far away in time and gives them a melancholic tone. It is the act of watching these films that allows viewers to "go for a walk," and that allows the city—that is, its people—to feel communally.[26] And while they are feature films, which are fictionalized accounts of the experience of the contemporary city, their modality attributes to them "truthfulness"—used here in the sense of veracity or genuineness and in a different manner from documentary realness.

Writings on media memories have touched upon the deliberate and relative uses of monochromatic and chromatic visuals in contemporary works, especially in cinema films.[27] In the realm of cinema, black-and-white footage is often associated with alternative reality: dreams, fantasies, memories, or the historical past. In art films, directors may use black-and-white or desaturated images for "truthfulness" rather than "reality" (e.g. Turkish director Nuri Bilge Ceylan, Russian director Andrei Tarkovski).[28] This distinction may at one level reflect the mid-twentieth century idea that most people dream in black-and-white.[29] Thus, the usage of monochromatic imagery alludes to the veracity of inner worlds—in opposition to the accurate, and perhaps more enjoyable, chromatic reproduction of the external world. Yet, the association of black-and-white with "pastness" or "truth" is a post-1960s phenomenon. It is in fact a reversal of the opposition that classical Hollywood had installed years earlier, when color connoted fantasy.[30] Although color photography was invented in the 1860s and color in films was used early as the 1920s, black-and-white photography and motion pictures (including news reels, cinema films, etc.) remained commonplace until the 1950s (and 1970s in Turkey). This led to the association of black-and-white with reality and color with fantasy: Most memorably, the black-and-white Hollywood classic *The Wizard of Oz* (1939) turns to color when its heroine Dorothy wakes up in the colored world of the Land of Oz. With high-definition color media saturating our everyday lives black-and-white has emerged as closer to "truth" today cast in opposition to the "reality" of color. Yet, black-and-white film was not a matter of choice but a standard of the film industry in Turkey until the 1970s. Only in the 1990s, when black-and-whiteness came to function as an aesthetic sign could old black-and-white films, remarkably deficient in technology compared to today's standards, make a comeback as memory objects.

Housing and belonging in the city

The films discussed here are not only black-and-white (monochrome) but also depict the city through a black-and-white vision of housing. The affluent middle classes live in modernist concrete-frame apartments, while the poor occupy decaying old wooden houses in inner city slums or poorly-built shacks in squatter settlements—with the latter group always, and somewhat despairingly, aspiring to the life of the former. This vision does not reflect the full complexity of housing now or then, but it serves as a trope that fits well with the rich/poor, modern/

traditional, male/female axes of their central plot lines. In fact, popular Turkish films of the era did not introduce such tropes; they merely capitalized on them.

Literary and cinematic constructions of the urban experience via housing tend to draw upon concrete processes but also to fictionalize them with moralistic messages. Some of the accounts, especially those with a nationalist orientation, are critical of the modern districts north of the Golden Horn and associate them with Westernization and decadence. For example, in his famous short novel/long story *Fatih-Harbiye* of 1931, Peyami Safa characterizes Istanbul as a dual city.[31] Fatih ("Conqueror") is an old Muslim neighborhood on the historic peninsula, surrounding the great mosque and *medrese*. Harbiye ("Military School") is a newer section, located north of the Golden Horn, built around a modern military school and inhabited by people who are too Westernized for their own good. For Safa, the differences between Fatih and Harbiye are East (Fatih) versus West (Harbiye); in other words, they are civilization differences. In his didactic novel of realist prose, the daughter of a liberal Muslim family is temporarily lured by the glitter and modernity of the Beyoğlu and Harbiye areas, only "to return to her senses" and her roots in Fatih. This dualistic association of the historic peninsula with conservative Muslim Istanbul and the areas north of the Golden Horn with non-Muslim and Europeanized groups, which had begun a century earlier, intensified from the mid-twentieth century onward with internal migration to the city and a rising demand for housing. However, the housing makeup of the city was more complex than such a dualistic vision.[32]

One response to the shortage of housing was spontaneously constructed informal gecekondu (squatter) settlements at the peripheries of the city; another developed by the middle classes was the "build-and-sell" strategy enabled by laws that permitted the fragmentation of ownership on a multistory apartment building on a single plot and the sale of units to different owners. Still, it is the contrasting vision of the historic peninsula of dilapidated historic homes versus new modern settlements to the north that still dominates the early 1960s films discussed here, while they continue to register the much more complex growth of the city via new types of housing projects and developments. These old films provide a unique perspective on how urban transformation was discussed and interpreted at the time (see Figure 3.3).

Experiencing the city through migrants' eyes

Birds of Exile is perhaps the most famous rural-to-urban migration film in Turkish film history.[33] It works within mainstream depictions of the figures of the migrant and the urbanite, the country and the city, aiming to reveal Turkey's social reality as the director sees it. Like most films of this genre, it neither portrays the journey nor contrasts the rural to the urban, but rather depicts the experiences of recent migrants once they are in Istanbul. A family of elderly parents, three grown sons (Kemal, Selim, and Murat), and a daughter (Fatma) arrives in Istanbul in search of fame and riches, but they lose everything, including the daughter, who drifts

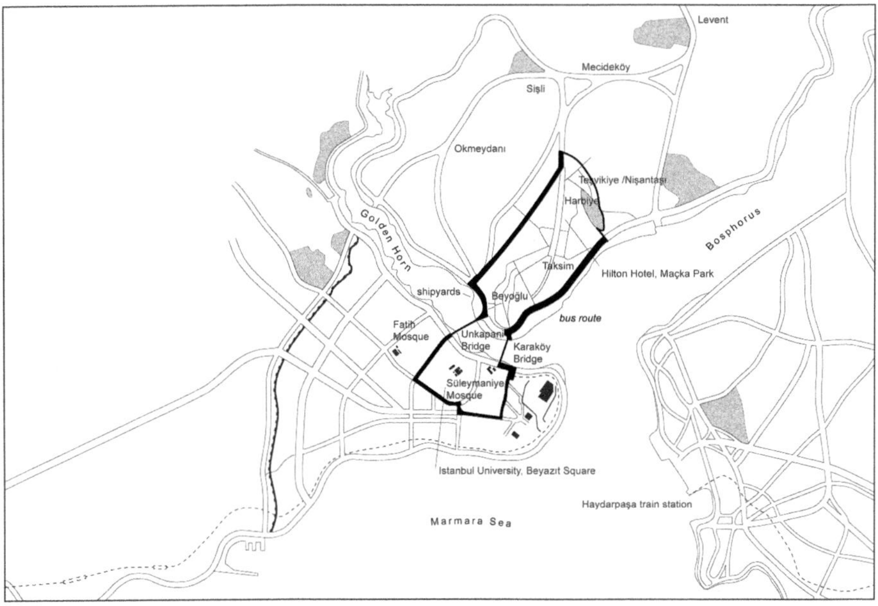

Figure 3.3 Diagrammatic map shows locations mentioned in the discussion of the three films.

Birds of Exile: Haydarpaşa Train Station, where the migrating family enters Istanbul; Beyoğlu and Nişantaşı/Şişli areas across the Golden Horn where the siblings' girl- or boy-friends live in concrete-frame apartment buildings; Hilton Hotel in Maçka Park.

Bitter Life: Haliç (Golden Horn) shipyards where the lead character works; Karaköy Bridge where he proposes to his partner; Okmeydanı where the couple visits new modernist housing, beyond their means.

Bus Passengers: The new bus route on the new boulevards depicted by Vedat Türkali in his memoir, which provided inspiration for the film; Süleymaniye Mosque which the lead hero, the bus driver, greets everyday; Istanbul University, where the lead heroine studies; Levent housing development, where scenes of new construction were shot; Mecidiyeköy, where Güvenevler buildings were built.

Map based on the new road network presented in *İstanbul'un* Kitabı (Istanbul: İstanbul Valiliği, 1957), and which excludes, among others, the two Bosphorus bridges and their highway connections, built much later.

into prostitution, and return home.[34] The director contrasts the failure of this relatively skilled and educated, aspiring migrant family in the city with the rise of the uneducated, opportunistic single male migrant, pointedly named "Haybeci" (freeloader), with whom they make the arrival trip to Istanbul at the beginning of the film as well as the departure trip at the end of it.

The entry sequence depicts the arrival of the migrant family in Istanbul via train at Haydarpaşa Train Station and follows their journey on a ferry across the Bosphorus, from Asia to Europe. Entry to Istanbul at Haydarpaşa Train Station is a trope of Turkish literature and cinema.[35] The director encourages identification with the members of the family and does so by contrasting them with Haybeci. Upon arrival, Haybeci claims his right to the city, first by avoiding paying for his train ride to Istanbul and for his boat ride across the waters of the Bosphorus; he then claims entitlement when confronted, arguing that his citizenship rights are rightfully gained due to the loss of family in various national wars. The family members observe from a distance and disapprove of this character.

The plot has some affinities with Luchino Visconti's *Rocco and His Brothers* (1960), but it also draws from journalistic reports on migrants; it reflects the director's anti-migration views, which in turn received coverage in the press through the film's reviews.[36] Refiğ's migrant characters initially display prowess participating in consumption: they join the urbanites in *flânerie*, strolling along the city's boulevards and in its parks, and socializing in its cafés, patisseries, theaters, and nightclubs. In this sense, the city provides a stage for leisurely exploration. The uncertainty of appearances is clearly pronounced in the romantic relationships the four siblings have. One of the siblings, Murat, falls in love with Naciye, an independent and self-willed woman who works in a *pavyon* (bar), believing she is an authentic Istanbulite. He is disappointed to find out that Naciye is also a migrant, in fact from his very own hometown, Maraş, in southeastern Turkey. Murat's younger brother, Kemal, who is a medical student, tries to pass himself off as an urbanite among his classmates. Kemal's Istanbulite classmate and girlfriend Ayla chides him for tipping a beggar on the street; she regards the beggar as an undeserving rural-to-urban migrant—unbeknownst to her is Kemal's real identity, revealed when they accidentally bump into his older brother Murat driving his cab. Such a comedy of errors repeatedly makes the point of the uncertainty of appearances.

Even if the siblings can hide and adjust their appearances, they cannot change their neighborhood. Even if public spaces enable encounters across social groups, social differences are marked by where people live.[37] Where the migrating family settles—in the historic peninsula, with its narrow, cobblestone roads, wooden houses, and decaying neighborhoods—is clearly a space for the urban poor.[38] It contrasts starkly with the city's affluent middle-class areas to the north of the Golden Horn, where the boy- or girlfriends of the siblings live. These are the northern, "European" parts of the city, with wide asphalt streets and modern concrete buildings. The area known as Beyoğlu/Pera, also to the north—with its disreputable entertainment establishments and churches, traditionally where the city's non-Muslim minorities lived—is home to the lovers of the two older sons, one the abovementioned bar-worker Naciye, and the other a Greek housewife/adulteress. Finally, there is another, "other," Istanbul, a place of squatter settlements, where rootless peasants take refuge. In its separation of Istanbul into two, *Birds of Exile* builds on a well-threaded image of the city, introduced before with *Fatih-Harbiye*, yet it also includes the squatter settlement in the map of the city.

Not only do the characters in *Birds of Exile* migrate and discuss migration, evoking at times their "right to the city" but also newspapers report on the many facets of the phenomenon, all within the diegesis. Viewing the film, we see how a sudden, sharp increase in the city's population was met by a rapid rise in housing development. We also see how the old wooden houses of the historic peninsula are being abandoned to a transitory lower middle class (which this particular migrating family represents), while the upper middle class is moving to modern concrete apartments in the northern part of the city. While the film demonstrates these transformations, it also reflects the anxieties engendered by urban modernity through new class encounters and consumption practices. One of the central concerns of the film is thus migration—migration to Istanbul by provincial skill-less or semi-skilled people, and immigration to Europe and the United States, via Istanbul, of educated young urbanites.[39]

Some of the most direct discussions of migration in *Birds of Exile* take place between the lead characters, Kemal and Ayla, both medical students. In a memorable sequence, shot in Maçka Park, they discuss their future. They decide to get married, but Ayla wants to go to America to study. During the discussion, the Istanbul Hilton Hotel (designed by the American firm SOM in collaboration with Turkish architect Sedad Eldem) is selectively placed between them in the frame. The use of this hotel, which was completed in 1955 and overlooks the city from its location in the park, is an important commentary on Turkey's political alliances at the time.[40] It is Americanization that stands between the two lovers, and by extension in the way of Turkey. By taking Ayla from here to the city's gecekondu settlements, Kemal convinces her not to pursue her graduate studies in the United States, but to stay in Istanbul and "mend [their] own home." The home Kemal shows Ayla is not Anatolia but the gecekondus in Istanbul.

Director Refiğ and his generation felt, in the aftermath of Turkey's first coup d'état in 1960, that the Turkish project of modernity had to be revised. They set out to give this project direction in serious journals such as *Yön* as well as through popular media. In *Birds of Exile*, Refiğ calls intellectuals to service. He depicts the city and its squatter settlements as the real problems faced by the nation (rather than the countryside). By substituting the city for the nation, he defines the citizen as an educated urbanite, suggesting that his/her commitments should not be to the countryside but to the city. In Refiğ's view of the city, the educated will remain, while the unqualified will go back where they came from. However, Refiğ also privileges morality in his criteria for a communitarian model of urban citizenship. His urbanite characters are judged in relation to the work they do for the city, and hence for the nation.

Ev-lenmek: Getting married (housed)

Bitter Life (1962) portrays a complementary image of the city through a working-class couple's eyes. Class differences are represented not so much through cultural, linguistic, or other visual subtleties but housing conditions. There have been many informal remakes of *Bitter Life* over the years. Most recently, in 2005, two popular TV dramas, one with the same title, were based on exactly the same topic.[41] Continuing press coverage of the original film shows it has persisted in public memory over the years.[42]

Director Erksan is cited generally as Turkey's first auteur director, marked by his Berlin Golden Bear award for *Susuz Yaz* (*Dry Summer*, 1964), and highly acclaimed takes on representation and gender, respectively in *Sevmek Zamanı* (*Time to Love*, 1965) and *Kadın Hamlet* (*Woman Hamlet*, 1977). In *Bitter Life*, he takes up a typical melodrama storyline, centered on a working-class couple's desire to get married and concomitant search for a place to live: The Turkish verb for getting married, *ev-lenmek*, literally means "acquiring a home." Since a place to live is a prerequisite, a good portion of the film depicts the couple, played by stars Ayhan Işık and Türkan Şoray, looking together in vain for suitable housing.

Following the title credits, the introduction sequence opens with frontal extreme close-ups of the faces of the two lead characters declaring their love to each other. The camera then quickly locates them in the city by zooming out a little and framing them together in a medium close-up, leaning on the banister of a Golden Horn bridge, with an Ottoman mosque (in the historic peninsula) in the background carefully placed between them. They decide to get married, to raise a family, and to look for a (rental) place to live in, symbolically taking their strength from the city. The camera then starts strolling along the crowded streets of Istanbul, assuming the couple's point of view. He accompanies her to the beauty parlor where she works and departs for his own work. She is a manicurist, who listens all day with visible distress to affluent customers' chatter as she tends to their nails and observes firsthand the modern spaces in which they live when she is called in for house service. In contrast, he is a welder at the docks. Socializing with similar working-class men engaged in productive labor, he is immune from monetary aspirations that come from exposure in the service economy.

She desires better living standards and modern amenities that are beyond the couple's means. They can only afford to live in a squatter shack or in one of the old, dilapidated wooden houses that serve as rooming houses for rural-to-urban migrants. Eventually, instead of her poor fiancé, she marries a rich suitor; but the ex-fiancé, by a twist of fate, wins the lottery, becomes a bitter businessman, and constructs a modern "villa" to show off to her. If Americanization comes in between the couple in *Birds of Exile*, it is a more stereotypical (and therefore universal) storyline in *Bitter Life*, where the gulf between the lovers is created by consumerist greed for wealth, manifested in modernist spaces and their modern amenities.

In one of the *Bitter Life*'s most memorable sequences, the couple is framed walking by the İETT (Istanbul Electricity, Tramway and Tunnel Company) blocks in Okmeydanı—built on land at the edge of the city that originally served as an archery practice ground for the Ottoman armies, but was turned over to developers in the 1950s. The multistory blocks under construction featured in this sequence were designed as residences for the İETT employees.[43] The couple happily walks by the smaller of these blocks, which clearly emulates the white egg-crate façade and horizontal massing of the Hilton Hotel. But the sight of the rental they are to view turns them pale. Across from the new blocks, all alone in the barren landscape, it is the second floor of a flimsy, unserviced timber shack. Disappointed, she wants to see the interiors of the new apartments. But a ten-second pan across the blocks' façades, from the couple's point of view, effectively communicates the burden the characters feel when they realize the financial impossibility of their spatial aspirations. In stark contrast to the opening sequence, which suggested the city's panoramic views as the essential backdrop for the formation of the lead characters' amorous bond, as they move from one apartment viewing to another, those feelings visibly weaken. By the end, love turns into vengeance and hate when both characters acquire access to wealth and status.

In *Birds of Exile*, when things do not work out in the city, the lead characters still have a home(town) to go back to. Yet, if they are rootless working-class Istanbulites, such as the characters in *Bitter Life* and *Bus Passengers*, resisting the temptations of the big city becomes more complicated. *Birds of Exile* and *Bitter Life* depict the city respectively through the eyes of migrants from Anatolia, and through the eyes of the working classes. *Bus Passengers*, in contrast, depicts it from those of intellectuals, proposing moving images of the city—literally as perceived through the windows of a public transportation bus in the film, but metaphorically speaking, via the cinematic apparatus—as a medium of resistance, and the space of class solidarity.

Moving images of the city

The eponymous bus of *Bus Passengers* follows a route that passes through the fictional neighborhood of Yeşiltepe, at the edge of the city, and ends at Beyazıt Square, in the heart of Istanbul. This is the location of Istanbul University, where the lead female character, Nevin (Türkan Şoray), studies. Kemal (Ayhan Işık), the bus driver, picks up the bus from its terminal each morning and starts his route at the Yeşiltepe stop, near which new concrete buildings abound. He stops at a quarry to drop off several workers and then continues south toward the historic peninsula, across the Unkapanı Bridge, and over the Golden Horn to his final destination, Beyazıt Square. Based on the look of the bus stop and its surroundings, the Yeşiltepe scenes appear to have been shot on location, in the brand-new, suburban settlement of Levent. The proximity in the film of Yeşiltepe to its nearby squatter settlement is also reminiscent of the condition of Levent as it appeared in the 1950s, alongside its squatter twin, Gültepe.

All the characters live on the periphery of Istanbul in distinct housing areas, but each day some travel on the bus all the way to the historic peninsula, to the center of the city. Three kinds of residential spaces dominate the film, each associated with a particular social group or class. Male workers in the quarry perform under dangerous conditions and constant pressure from despotic overseers. They live in tents without basic amenities and squat on the ground to eat their meals. The diegetic music, performed by a folk poet/worker, situates these workers as migrants of Anatolian origin. In the soundtrack, the authenticity of this music contrasts the artifice of (also diegetic) classical music that flows from the phonograph and radio of Feslioğlu, the developer, at his dinner table. Feslioğlu lives with his daughter, Nevin, his son, and his two sisters in a detached suburban villa. Despite the physical proximity of the villa to the squatter settlement, his family maintains social distance by spending most of their time in interior spaces and, when outdoors, driving in their private cars. Their villa, which is cluttered with household appliances and furniture, displays an ideal of bourgeois domesticity. In contrast, the squatter settlements are minimally decorated on the inside, lacking public spaces (cafés, squares, playgrounds) or even basic services such as paved streets, water, electricity, and public transport. Nevertheless, the residents spread out on the streets, and the portrayal suggests warm social relations and a cohesive community.

Every morning, members of each group come together on the public bus that takes them to the center of the city. Arguably, the physical expansion and transformation of the city turns the bus ride into the only remaining way to engage with it, from a distance and within a moving apparatus. The bus ride provides moving images of the city, clearly articulated as such by its driver, the hero. Indeed, the bus itself is cast by the filmmakers as a new, idealized public space.[44] As will become clear, the bus is also a figurative substitute for cinema.

The film has two parallel stories: a love affair, and a community struggle around real estate speculation and corruption. The love story reproduces many of the narrative conventions of Turkish melodrama, particularly the integration of vistas of the city into the plot. The bus driver Kemal (played by Işık) falls in love with Nevin. The two grow increasingly fond of each other as they visit canonical sights around the city. But their outings soon lead to rumors, and their class backgrounds get in the way. Typical of its genre, however, these narrative tensions are resolved toward the end of the film, and it concludes with the couple's happy union.

The second, but more important, story addresses social groups and processes that were becoming increasingly visible at the time due to the transformation of the city. In the film, many of the people living in the squatter community aspire to one day move into the modern flats being produced by a housing cooperative. Many invest their life savings for this purpose. However, the film uses imagery reminiscent of the ads in the daily press (see Figures 3.4, 3.5) to expose their true

Figure 3.4 "The Opportunity to Live Comfortably." Newspaper ad (for a lottery prize apartment) for a bank urging readers to abandon their wooden historic houses for concrete frame apartment buildings.

Ad by Türkiye İş Bankası (Business Bank of Turkey), "Daha Rahat Yaşamak İmkanı," *Yeni İstanbul*, September 22, 1958.

Figure 3.5 "The Desire of Thousands of People." Clerks in the showroom of the Güvenevler housing cooperative in the film *Bus Passengers* mimic gestures and framing in ads (again, for lottery prize apartments) as shown here.

Ad by Yapı ve Kredi Bankası (Building and Credit Bank), "Binlerce Kişinin Arzusu," *Hürriyet*, July 6, 1955.

intent—to entice uninformed buyers to purchase a product, at any cost. In the showroom of the construction company, the sales agents use same tactics as in the ads (claiming that a modern flat is "the desire of thousands of people"). They present the new housing as an object that will fulfill the buyer's dream of modernity. In contrast to the "American Dream" attached to the suburban house of the Levittown type, with its associated ideals of a horizontal society, the "(Turkish) Dream" presented here is about verticality associated with social upward mobility and achieved through concrete-frame walk-ups.

The buyers, or shareholders, are devastated during the course of the film to learn that the whole cooperative has been a fraud. They discover that each of its flats has been resold multiple times. Not only this, but other legal and structural problems plague the project. Meanwhile, the land speculator, developer, and a local politician continue to collaborate to maximize their profits, and the film shows how their partnership privileges personal profit over civic duty or professional ethics.

Eventually, the members of the bus community call on Kemal to defend their rights. Although at first reluctant to became engaged in social or political issues, Kemal quickly embraces his new role. Donning his official cap and tie, he takes the developer to court and writes letters to all the newspapers. The story makes front-page news. Alarmed, the politician terminates his contact with the speculator and the developer, who quickly dissolve their partnership after recognizing it has started to crumble.

In a sense, Kemal stands in for the director on one level and public intellectuals on another. He encourages the community to organize itself to demand its rights, use the legal process, and involve the press to generate a wave of public opinion. Public opinion, in return, becomes corrective of the state. Ultimately, the people "fight back" and triumph over the developer, the land speculator, and their hired agents. The state, embodied by the police, arrives to arrest the greedy developer and speculator and protect them from public lynching. When read in the political context of the times, the film suggests the right to housing as the basis of political consciousness and the emergence of social movements.

Fictionalization: from real to reel

Reflective of the organization of the film industry, these films fictionalized accounts in journalistic reports and were shot on location, displaying a remarkable instantaneity in a pre-TV era.[45] They turned the city into a readily viewable object of consumption. I focus here on *Bus Passengers* to demonstrate and reflect upon the degree of fictionalization of real-life events.

In interviews and memoirs, Türkali and Göreç explain that there were multiple sources of inspiration for *Bus Passengers*. The first was the urban-renewal operations the city was undergoing. In the latter half of the 1950s, new boulevards were cut through the historic fabric. These isolated monumental sites and offered new vistas. The İETT acquired new buses to travel on these boulevards. Second, the vibrant, speculative housing sector was producing a new social hierarchy. It was also creating conflicts, which would at times produce litigation between developers, aspiring homeowners, professionals, and state authorities. Türkali and Göreç examined the records of one such incident, the Güvenevler (Trust Homes) case, to better inform a subplot in the film. Third, the general political atmosphere in Turkey encouraged filmmakers to deal with the pressing urban issues of the time. Türkali summarizes the various sources of inspiration for the film as follows:

> The new public buses that arrived in Istanbul made a ring taking off from the corner of today's Ankara Pazarı in Nişantaşı, serviced the recently opened

> Unkapanı Bridge, Beyazıt via Aksaray, and returning via Sirkeci, Karaköy, Dolmabahçe, and Maçka. In those days, these buses, which ran on the new roads that revealed the stunning beauties of Istanbul, brought about a new air to the city … . During those days, a swindle by a partnership named Trust Homes on the periphery of the city appeared in the press. The administrators were charged and taken to court. What we would tell would be the exploits of the corrupt building partners who exploited the aspirations of the people, the middle layers, and their aspirations of a roof over their heads … . The bus driver and the attractive girl at the front of the vehicle suited this story.[46]

The newly acquired, red, flat-front buses stirred up genuine excitement because of their design, technology, and speed. Motor vehicles, including private cars and public buses, were still scarce in Istanbul, and traffic jams and minor accidents caused by novice drivers easily made the front pages of national newspapers. As late as 1956 no more than 157 buses had been operational. But during the urban renewal of 1956–60, 428 new buses were imported and put into service on the new boulevards.[47] Until then the dominant form of public transport had been the tramway, but rails for the trams were removed as part of the urban renewal program. This process began in Aksaray with the opening of the Avenue of the Nation (Millet Caddesi), and, after a public farewell, the trams were completely removed from the European side by 1961.[48]

The buses that were acquired to replace the trams signified a new era. Their bright red color, relative speed, and maneuvering ability turned them into everyday icons of modernity. Unlike the tram that preceded it, the bus did not divide passengers into first and second class. The public buses remained among the few spaces in which different social groups could come together in a city increasingly segregated by new residential patterns and consumption practices. But most importantly, the bus offered an elevated and collective view of the city in motion. Thus, it was only appropriate that *Bus Passengers* called attention to it as an important new public space. The film further aimed to revise class-based agonistic encounters by portraying the possibility of convivial and lively debate within the community of the bus.

Türkali and Göreç first became aware of the Trust Homes property fraud because this particular project was marketed in cinema journals and many of their acquaintances were involved as buyers or cooperative members. The case of Trust Homes was also unusual because it made its way onto the front pages of national dailies (see Figures 3.6, 3.7).[49] A print ad for Trust Homes, published in a cinema newsletter at the end of the 1950s, promised: "If you wish to own a comfortable flat with long term mortgage, the Trust Homes Block Housing Collective Incline which has made 1,000 families homeowners in 19 months [is] in your service." The image in the ad depicts an austere, tall block building next to a three-story, glass-clad lower block building. The title of the project, "Güvenevler," is drawn over the lower block in building-size, tall letters that emphasize the importance of its brand. In reality, there was not much to be trusted about this cooperative. To begin, this project in Mecidiyeköy was not a cooperative at all but a typical large-scale

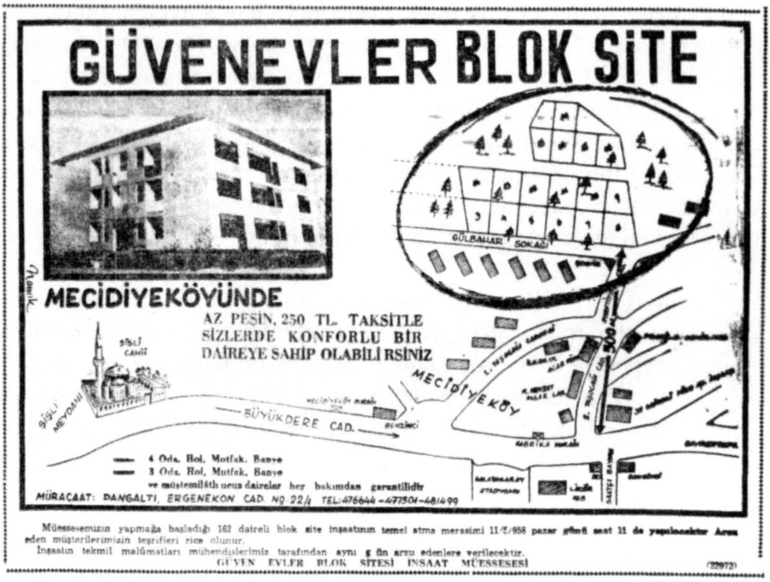

Figure 3.6 Ad for Güvenevler.

Cumhuriyet, September 5, 1958, 6.

speculative development similar to others that had started to appear on the periphery of the city. It merely disguised itself as a cooperative to be eligible for state-subsidized low-interest loans.

Trust Homes' filmic interpretation, *Bus Passengers*, portrays the residents of the squatter settlement as future users of planned, modern apartment buildings. Yet, even though the construction of these buildings continues throughout the film, there is no indication that any of the residents will ever move in. The film thus questions the belief promoted by modernization theory that squatters eventually become urbanites, and thus, in the case of Turkey, that present substandard housing represents a transitory stage in the country's modernization. The construction in the film therefore doubles as the stage set of modernity; but, as such, it remains an unfulfilled desire. By contrast, the visual consumption of the city, especially through the public bus, offers a liberating alternative.

Bus Passengers deliberately critiqued the new social groups and values that emerged as a result of housing consumption. Delinking the opening of boulevards and housing consumption, it interpreted the boulevards as an opportunity. In this sense, the public bus doubled as a "moving image" of the city, facilitating a new public realm for different social groups to gather in. However, even if the film reflected belief in the potential of the newly expanded public sphere to create new types of communities, it was suspicious that these new communities would be able to act in their common interest. Its view was that the intellectual must step up and do this. Here, the associations are evident; cinema as apparatus parallels

Dünkü kongrede heyecanlanan üyeler polisin yardımı ile teskin ediliyor

Güven Evler Yapı Kooperatifi ortakları 200 hane ve 4 milyon 484 bin lira alacaklı durumda

Kooperatifin dünkü olağanüstü kongresi gürültülü bir hava içinde geçti ve üyeler hâlen tutuklu bulunan kooperatif kurucuları hakkında ağır sözler sarfettiler

Güvenevler Yapı Kooperatifinin olağanüstü kongresi, dün gürültülü bir hava içinde yapılmıştır.

Kooperatifin tasfiyesi için kanuni ekseriyetin sağlanamadığı bu kongreyi, alkışlar arasında başkanlığa seçilen, Rafet Kaya adındaki bir albay ile yine subaylardan teşekkül etmiş bir kurul idare etmiştir.

İki ay önce seçilen idare kurulundan kalan üç üyeden Vehbi Coşkun, yaptığı konuşmada, Faik

—Arkası Sa. 5, Sü. 3 te —

Figure 3.7 Güvenevler cooperative dispute received wide coverage in the newspapers of the time. Shown here, the dispute in a cooperative meeting is calmed by the police.

Cumhuriyet, July 25, 1960, 1.

the role ascribed to the bus, and the filmmaker parallels that of the intellectual/bus driver.

The bus driver, as imagined in the film, displays a genuine fondness toward the new openness of the city. The film explains, through diegetic, lengthy self-narration, that Kemal aestheticizes the city as a counterbalance to the lack of social justice. Every morning, Kemal greets the Süleymaniye Mosque complex with passion. In fact, the picture of Süleymaniye is framed on his bedroom wall—a popular practice at a time when illustrated magazines provided foldout posters of Istanbul's sights and instructed readers to hang them on their walls. It is his appreciation of the city's architecture, Kemal explains, that enables him to "stand against all the wrongdoing, all the ugliness."[50] It also transforms his manual labor, driving a bus on the city's new boulevards, into a noble one. Thus, the film promotes enjoyment of the city as an image within both the space of the film and the space of the theater. In lieu of other forms of material consumption, it thereby suggests that this mode of aesthetic engagement with the city may make it possible to transcend the trap of passive consumerism, and become the basis even of solidarity, and political change.

At the time of its release, *Bus Passengers* provided an oblique assessment of the rule of the Democrat Party, and its making was only possible after the overthrow of that government in a military coup in 1960. In particular, the film critiques power relations within the construction industry, which, as one of the leading sectors of the urban economy, seems to have created its own social hierarchy. The film aims to expose the corruption of capitalist relations—in particular, how modest people living on the margins of the city have been conned by a coalition of politicians, capitalists, and technocrats with promises of modernity through housing. It points to the inability of a group of cooperative housing victims to organize on their own to press for their rights. And it proposes an alternative form of belonging, marked by a cultured enjoyment of the city through images—i.e. that offers the cinematic city in lieu of the real one.

Cinema and the memory of urban experience

The revalorization of old Turkish cinema films is partially connected to the recognition that cosmopolitan origins and international connections of cinema continued to shape Istanbul's cultural landscape well into the Republican era and, to a limited extent, to the present day. Istanbul was the locus of cinema production—production houses, artists' agencies, and large film theatres all concentrated in Beyoğlu/Pera, the historic neighborhood of Levantines and non-Muslims. Just like Pera, cinema was a multicultural enterprise, despite the nationalistic overtones typical of the films. Yet, it is not possible to read this from film plots.

The recirculation of old films may partially have to do with the availability of media space and the desire to fill it with familiar and cheap content. Before the late 1990s, it would have been embarrassing to admit to being a fan of these old films. More typically, they were mocked for their deliberate excess—eliciting laughter or tears based on unrealistic rags-to-riches stories.[51] Their citation and quotation, however, suggests an anxiety that has to do with Istanbul's integration into global markets, the "fragmentation" and privatization of all aspects of life, which raise the need for media memories to fashion alternative selves.

Istanbul was a central character in many old Turkish films—most of them were shot in Istanbul on location and were based loosely on current news reports that had to do with Istanbul's demographic growth, urban expansion, renewal, and housing conditions. Old Turkish films are valuable to examine today not only because they have documentary value but also because they have transformed into memory objects as they circulate in contemporary visual culture.

These films, I have suggested, provide black-and-white visions of the city. Black-and-white functions in a specific way today and is associated with veracity—hence the warm reception of old films. I have also used "black-and-white" as a metaphor for the portrayal of class struggles and the associated housing make-up of the city. As depicted in these films, the middle classes typically live in modern, concrete-frame apartment buildings and the poor live elsewhere, in inner-city wooden houses that have turned into slums, or squatter settlements on the periphery of the city.

Yet the real situation was not, and has never been, as clear-cut. Not all residents of squatter settlements were recent rural-to-urban migrants without skills; some were educated and had merely been evicted due to urban renewal and demolitions or displaced due to economic hardship. Even the apartment building, which had a middle-class association, had more complexity in practice. The higher the unit's floor, the more desirable or prestigious it would be. Apartment buildings were never exclusive to the middle classes; female domestic helpers typically worked in them, and a key feature of these buildings was the tiny "kapıcı" (doorman) unit in the basement, occupied by a recent rural-to-urban family. The three films I have focused on recycle conventional class/housing dichotomies but also portray some of these complexities. In addition, these films deliberately use specific shots of the city, its vistas and panoramas, to vocalize their political perspectives on citizenship rights as well as duties. The vistas of the city empower the films' characters and, via the apparatus, the films' viewers.

Today's critiques are directed at urban sprawl and sustainability, or the lack thereof, in new developments, as depicted in the documentary *Ecumenopolis: The City Without Limits* (2012, dir. İmre Azem) as well as much of academic writing on "urban transformation." The giant state apparatus for privatization, TOKİ, under the AKP since 2002, oversees the sprawl of not only Istanbul but all Turkish cities with standard high-rise apartment blocks of a particular aesthetic, dispossessing poor and marginal people from both inner city slums and informal settlements on the fringes of the city. It is important to remember that the criticism directed at the AKP applies to previous governments such as Turgut Özal's Anavatan (Motherland) Party (1983–93) and Adnan Menderes' Democrat Party. Its scale and objectives, actors and perceived issues, are much transformed, but housing production remains the main economic engine of the urban economy today.

If *apartmanlaşma* marked the post-1950 decades of housing experience for the middle classes, from the 1990s on, they further segregated themselves in gated communities.[52] Squatter settlements of rural-to-urban migrants and the urban poor, which were the locus of clientelist politics—and therefore the focus of sociological studies on the city—have been legalized and incorporated in the formal economy since the 1980s. The three films discussed here present the association of *apartman* with middle-class urbanite status and *apartmanlaşma* with modernization—an association that clearly wears away in the mid-1970s when a majority of Istanbulites had access to apartments, and is marked by the comeback of old historic homes in exhibitions on the city—calling for their preservation as part of the urban fabric (as discussed in Chapter 4).

Notes

1 Scott McQuire, *The Media City: Media, Architecture, and Urban Space* (London: Sage, 2008). On "urban imaginary," see Alev Çınar and Thomas Bender, "Introduction: The City, Experience, Imagination and Place," in *Urban Imaginaries: Locating the Modern City*, ed. Alev Çınar and Thomas Bender (Minneapolis, MN: University of Minnesota Press, 2007), xi–xxvi.

2 David B. Clarke, ed., *The Cinematic City* (London: Routledge, 1997); Nezar AlSayyad, *Cinematic Urbanism: A History of the Modern from Reel to Real* (New York: Routledge, 2006); Barbara Mennel, *Cities and Cinema* (London: Routledge, 2008). Some of the other publications that interrogate the relationship between the city and cinema include François Penz and Maureen Thomas, eds. *Cinema & Architecture: Méliès, Mallet-Stevens, Multimedia* (London: British Film Institute, 1997); and Mark Shiel and Tony Fitzmaurice, eds. *Cinema and the City: Film and Urban Societies in a Global Context, Studies in Urban and Social Change* (Oxford; Malden, MA: Blackwell, 2001).

3 Jean Baudrillard, *America*, trans. Chris Turner (London: Verso, 1988), 56.

4 Georg Simmel, "The Metropolis and Mental Life (1903)," in *Simmel on Culture: Selected Writings*, eds. David Frisby and Mike Featherstone (London; Thousand Oaks, CA: Sage, 1997), 174–85. Along the same lines, Siegfried Kracauer's *Salaried Masses* (first published in 1930) provides a classic account of the role of distraction and entertainment in the modern metropolis. Siegfried Kracauer, *Salaried Masses: Duty and Distraction in Weimar Germany*, trans. Quintin Hoare, intro. Inka Mülder-Bach (London: Verso, 1998).

5 Walter Benjamin, "The Work of Art in the Age of Its Technological Reproducibility (1936 to 1939)," in *The Work of Art in the Age of Its Technological Reproducibility, and Other Writings on Media*, eds. Michael W. Jennings, Brigid Doherty, and Thomas Y. Levin, trans. Edmund Jephcott, Rodney Livingstone, Howard Eiland, and others. (Cambridge, MA: The Belknap Press of Harvard University Press, 2008), 19–55; Walter Benjamin, "Theory of Distraction," in *The Work of Art in the Age of Its Technological Reproducibility*, 56–7; Howard Eiland, "Reception in Distraction," *Boundary* 2, no. 30 (2003): 51–66.

6 Tony Fitzmaurice, "Film and Urban Societies in a Global Context," in *Cinema and the City: Film and Urban Societies in a Global Context*, eds. Mark Shiel and Tony Fitzmaurice (Oxford; Malden, MA: Blackwell, 2001), 28.

7 On the connection of Turkish cinema film industry and the post-1980 era TV series, see: Eylem Yanardağoğlu, "TV Series and the City: Istanbul as a Market for Local Dreams and Transnational Fantasies," in *Whose City Is That? Culture, Design, Spectacle and Capital in Istanbul*, eds. Dilek Özhan Koçak and Orhan Kemal (Newcastle: Cambridge Scholars' Publishing, 2014), 47–63. On the roots of new popular cinema in the popular Turkish films of the 1950s and 1960s, see: Catherine Simpson, "Turkish Cinema's Resurgence: The 'Deep Nation' Unravels," *Senses of Cinema* 39 (May 2006), accessed April 12, 2010, http://sensesofcinema.com/2006/feature-articles/turkish_cinema.

8 Agah Özgüç, *Türk Sinemasında İstanbul* (Istanbul: Horizon International, 2010); Semra Kır, *İstanbul'un 100 Filmi* (İstanbul Büyükşehir Belediyesi Kültür A.Ş. Yayınları, 2000).

9 These films are fully available online, streaming via YouTube. Accessed on January 18, 2017. *Birds of Exile*: https://www.youtube.com/watch?v=x8SQp4ySQ0k. *Bitter Life*: https://www.youtube.com/watch?v=kHBHZkcblqo. *Bus Passengers*: https://www.youtube.com/watch?v=cadQKrKu528.

10 Nezih Erdoğan and Deniz Göktürk, "Turkish Cinema," in *Companion Encyclopedia of Middle Eastern and North African Film*, ed. Oliver Leaman (London: Routledge, 2001), 537.

11 Here, I am inspired by Miriam Hansen, who suggests that the public sphere is an appropriate concept to think about (early) cinema. Miriam Hansen, "Early Cinema, Late Cinema: Transformations of the Public Sphere," in *Viewing Positions: Ways of Seeing Film*, ed. Linda Williams (New Brunswick, NJ: Rutgers University Press, 1995), 144, 145.

12 Intellectuals, who were disenchanted with Yeşilçam or Hollywood films, but also believed in the potential of a more refined film taste, set up their own alternative screenings and venues for arthouse films. The longest surviving and most popular of these clubs was the Sinematek (Cinematheque, 1965–80), with its membership reaching 10,000 during its first few years of operation. Zeynep Avcı, "İstanbul Film Festivali'ne Ulaşan Yol: Onat Kutlar ve Şakir Eczacıbaşı, Sinematek Dönemini Anlatıyor," *İstanbul*, no. 9 (1994): 147–54.

13 Nilgün Abisel, *Türk Sineması Üzerine Yazılar* (Ankara: Phoenix, 2005), 200.

14 Arzu Kalemci and Şükrü Özen, "Institutional Change in the Turkish Film Industry (1950–2006): The 'Social Exclusion' Impact of Globalization," *TODAIE's Review of Public Administration* 5, no. 1 (March 2011): 69–120.

15 Nezih Coş, "Türkiye'de Sinemaların dağılışı," *As Akademik Sinema*, no. 2 (1969): 19–20; and Nezih Coş, "İstanbul'un Sinemaları…" *As Akademik Sinema*, no. 4 (1969): 11–20.

16 Giovanni Scognamillo, *Türk Sinema Tarihi, 1896–1997* (Istanbul: Kabalcı, 1998).

17 Dilruba Çatalbaş, "Broadcasting Deregulation in Turkey: Uniformity within Diversity," in *Media Organisations in Society*, ed. James Curran (London: Arnold, 2000), 126–48; and Catherine Simpson, "Turkish Cinema's Resurgence: The 'Deep Nation' Unravels."

18 Deniz Göktürk, "Turkish Delight–German Fright: Migrant Identities in Transnational Cinema," in *Mediated Identities*, eds. Karen Ross, Deniz Derman, and Nevena Dakovic (Istanbul: Bilgi University Press, 2001), 131–49; Deniz Göktürk, "Beyond Paternalism: Turkish German Traffic in Cinema," in *The German Cinema Book*, eds. Tim Bergfelder, Erica Carter, and Deniz Göktürk (London: BFI, 2002), 248–56; and Gönül Dönmez-Colin, *Turkish Cinema: Identity, Distance and Belonging* (London: Reaktion, 2008), 216–18.

19 The first annual conference on Turkish film research, Türk Film Araştırmalarında Yeni Yönelimler (New Directions in Turkish Cinema), was held in Istanbul in 1999 at a private university. There has been a deliberate interest within Turkish film studies circles since then to analyze early films, and a small but increasing effort has been made to connect cinema and the city (proceedings, Bayrakdar, 2001–2008). Some local publications that take on cinema and the city include Mehmet Öztürk, *Sine-Masal Kentler: Sinematografik Bir Üretim Alanı Olarak Kent Üzerine Bir İnceleme* (Istanbul: Om, 2002); and Nurçay Türkoğlu, Mehmet Öztürk, and Göksel Aymaz, eds., *Kentte Sinema, Sinemada Kent* (Istanbul: Yeni Hayat Yayıncılık, 2004). Earlier works that concentrate on representations of migration to the city are Oğuz Makal, *Sinemada Yedinci Adam: Türk Sinemasında İç ve Dış Göç Olayı* (Izmir: Köprü, Mars Matbaası, 1987); and Gülseren Güçhan, *Toplumsal Değişme ve Türk sineması: Kente Göç Eden İnsanın Türk Sinemasında Değişen Profili* (Ankara: İmge Kitabevi, 1992).

20 Nezih Erdoğan, "Narratives of Resistance: National identity and ambivalence in the Turkish melodrama between 1965 and 1975," in *Asian Cinemas: A Reader and Guide*, eds. Dimitris Eleftheriotis and Gary Needham (Honolulu: University of Hawaii Press, 2006) [original edition: *Screen* 39, no. 3 (1998): 259–71]; and Deniz Bayrakdar, "Türk Sineması: Hayali Vatanımız?" in *Türk Film Araştırmalarında Yeni Yaklaşımlar IV*, ed. Deniz Bayrakdar (Istanbul: Bağlam Yayınları, 2004).

21 Giuliana Bruno, *Atlas of Emotion: Journeys in Art, Architecture and Film* (London: Verso Press, 2002), 254.

22 Asuman Suner, *Haunted House: Belonging, Identity, and Memory in New Turkish Cinema* (New York: I.B. Tauris, 2010), 20.

23 Alison Landsberg, *Prosthetic Memory: The Transformation of American Remembrance in the Age of Mass Culture* (New York: Columbia University Press, 2004).

24 Orhan Pamuk, *Istanbul: Memories of a City*, trans. M. Freely (London: Faber and Faber, 2005), 32–3.
25 Orhan Pamuk, Foreword to *Ara Güler's Istanbul* (London: Thames and Hudson, 2009).
26 Pamuk, *Istanbul*, 95.
27 Paul Grainge, ed., *Memory and Popular Film* (Manchester, UK; New York: Manchester University Press, 2003).
28 The distinction belongs to art critic Anne Hollander in *Moving Pictures* (New York: Alfred A. Knopf, 1989), 33: "photographs and movies in black and white are considered good because they are so true, not because they are so real. . . . 'living color' may be more lifelike and more delicious, but, like life itself, it is also more distracting, entrancing and misleading." Quoted in Vida T. Johnson and Graham Petrie, *The Films of Andrei Tarkovsky: A Visual Fugue* (Bloomington; Indianapolis, IN: Indiana University Press, 1994), 189.
29 Eric Schwitzgebel, "Why Did We Think We Dreamed in Black and White?" *Studies in History and Philosophy of Science*, no. 33 (2002): 649–60.
30 Richard Misek, *Chromatic Cinema: A History of Screen Color* (Chichester, UK: Wiley-Blackwell, 2010).
31 Peyami Safa, *Fatih-Harbiye* (Istanbul: Ötüken Nesriyat, 1978).
32 Çağlar Keyder, "The Housing Market from Informal to Global," in *Istanbul: Between the Global and the Local*, ed. Çağlar Keyder (Lanham, MD: Rowman & Littlefield, 1999), 149. See also: Uğur Tanyeli, *İstanbul 1900–2000: Konutu ve Modernleşmeyi Metropolden Okumak* (Istanbul: Akın Nalça, 2004); Sibel Bozdoğan and Esra Akcan, "Housing in the Metropolis," *Modern Architectures in History: Turkey* (London: Reaktion, 2012), 139–69.
33 Ipek Türeli, "Istanbul through Migrants' Eyes," in *Orienting Istanbul: Cultural Capital of Europe?*, eds. Deniz Göktürk, Levent Sosyal, and Ipek Türeli (New York; London: Routledge, 2010), 144–64.
34 To be more precise, she is encouraged to jump off the roof of an apartment building to her death in fear and in full view of her enraged brothers.
35 It appears for the first time in cinema in 1950 in the aforementioned *Istanbul Nights* (1950). Most notably, Hatice Ayten's short documentary film *Menschen auf der Treppe* (*People on the Stairs*, 1999) depicts the experience of migration with a focus on the history of Istanbul's two main rail stations, Sirkeci and Haydarpaşa, and makes the point that Haydarpaşa appears in a wide range of artistic works, from poetry to feature films. Feride Çicekoğlu, "Threshold of the City: Haydarpaşa Train Station," in *World Film Locations: Istanbul*, ed. Özlem Köksal (UK: Intellect: 2012), 42.
36 "Taşradan Büyük Şehre Göç: Bir Rejisor İstanbul'a Akın Meselesini Yeni Filmine Konu Olarak Ele Aldı," *Milliyet*, December 23, 1963, 6; and Doğan Pursun, "Göç konusu Beyaz Perdeye Aktarıldı," *Akşam*, January 5, 1964.
37 For a more detailed discussion, see: Türeli, "Istanbul through Migrants' Eyes."
38 The relevant sequence from the film is viewable at: https://www.youtube.com/watch?v=L_oM_DEGLMA&feature=player_embedded.
39 The trope of migration has been especially prominent in Turkey in popular genres such as melodrama and comedy. Even a contemporary art-house film such Nuri Bilge Ceylan's *Distant* (*Uzak*, 2003) can focus on it.
40. Annabel Jane Wharton, *Building the Cold War: Hilton International Hotels and Modern Architecture* (Chicago, IL: University of Chicago Press, 2001).
41 Bülent İpek, "Erksan'ın 'Acı Hayat'ı İki Diziye Konu Oldu," *Sabah*, October 27, 2005.
42 "İlgi Çekici bir Türk Filmi: Acı Hayat," *Milliyet*, March 31, 1963; "Manikürcü kız, Türkan Şoray'ın kaderini çizdi," *Milliyet*, June 6, 1985; and "Sınıf Atlama Üzerine bir Melodram," *Milliyet*, August 12, 2001.
43 The sequence is viewable at: https://www.youtube.com/watch?v=Cblrxje9zn4&feature=player_embedded.

44 The conviviality of the bus ride is depicted in the following sequence: https://www.youtube.com/watch?v=exDPak4Ezxo&feature=player_embedded.
45 Serpil Kırel, *Yeşilcam Öykü Sineması* (Istanbul: Babil, 2005).
46 Author's translation. Vedat Türkali, *Eski Filmler* (Istanbul: Gendaş Kültür, 2003).
47 *İstanbul Ulaşımında 50. Yıl* (Ankara: Karayolları Genel Müd., 1974), 121–2.
48 The reasons for the retirement of the trams included low speeds, their inability to maneuver, a tendency to block motor traffic, high maintenance costs, and relatively low levels of comfort for the passengers. Sertaç Kayserilioğlu, *Dersaadet'ten İstanbul'a Tramvay* (Istanbul: İETT Genel Müdürlüğü, 1998).
49 In *Akşam* alone, the following news reports appeared: "Güvenevlerin yaptığı inşaat yıktırılacak," *Akşam*, April 23, 1960, 1; Hasan Bedrettin Ülgen, "Güvenevler mahallesinde halk dert yanıyor," *Akşam*, April 24 1960; "Güvenevlerin arazisine Belediye sahip çıkıyor," *Akşam*, May 1, 1960, 2; "Güvenevler inşaati meselesi imar vekaletince ele alındı," *Akşam*, May 8, 1960, 2; "Güvenevler dolandırıcı sahipleri ortadan kayboldu," *Akşam*, June 19, 1960, 1.
50 Türkali, *Eski Filmler*, 74–5.
51 Bilge Ebiri, "How Does It Feel to Feel? Recent Turkish Cinema," *Cinema Scope: Expanding the Frame on International Cinema* 23 (2005), accessed February 28, 2010, http://www.cinema-scope.com/cs23/fea_ebiri_turkish.htm.
52 Jean-François Pérouse and A. Didem Danış, "Zenginliğin Mekânda Yeni Yansımaları: İstanbul'da Güvenlikli Siteler," *Toplum ve Bilim*, no. 104 (2005): 92–123.

Bibliography

Abisel, Nilgün. *Türk Sineması Üzerine Yazılar*. Ankara: Phoenix, 2005.

AlSayyad, Nezar. *Cinematic Urbanism: A History of the Modern from Reel to Real*. New York: Routledge, 2006.

Avcı, Zeynep. "İstanbul Film Festivali'ne Ulaşan Yol: Onat Kutlar ve Şakir Eczacıbaşı, Sinematek Dönemini Anlatıyor." *Istanbul*, no. 9 (1994): 147–54.

Baudrillard, Jean. *America*, translated by Chris Turner. London: Verso, 1988.

Bayrakdar, Deniz. "Türk Sineması: Hayali Vatanımız?" In *Türk Film Araştırmalarında Yeni Yaklaşımlar IV*, edited by Deniz Bayrakdar. Istanbul: Bağlam Yayınları, 2006.

Benjamin, Walter. "The Work of Art in the Age of Its Technological Reproducibility (1936 to 1939)." In *The Work of Art in the Age of Its Technological Reproducibility, and Other Writings on Media*, edited by Michael W. Jennings, Brigid Doherty, and Thomas Y. Levin, translated by Edmund Jephcott, Rodney Livingstone, Howard Eiland, and others, 19–55. Cambridge, MA: The Belknap Press of Harvard University Press, 2008.

—. "Theory of Distraction." In *The Work of Art in the Age of Its Technological Reproducibility, and Other Writings on Media*, edited by Michael W. Jennings, Brigid Doherty, and Thomas Y. Levin, translated by Edmund Jephcott, Rodney Livingstone, Howard Eiland, and others, 56–7. Cambridge, MA: The Belknap Press of Harvard University Press, 2008.

Bozdoğan, Sibel, and Esra Akcan. *Modern Architectures in History: Turkey*, London: Reaktion, 2012.

Bruno, Giuliana. *Atlas of Emotion: Journeys in Art, Architecture and Film*. London: Verso Press, 2002.

Çatalbaş, Dilruba. "Broadcasting Deregulation in Turkey: Uniformity within Diversity." In *Media Organisations in Society*, edited by James Curran, 126–48. London: Arnold, 2000.

Çicekoğlu, Feride. "Threshold of the City: Haydarpaşa Train Station." In *World Film Locations: Istanbul*, edited by Özlem Köksal. UK: Intellect: 2012.
Çınar, Alev, and Thomas Bender, eds. *Urban Imaginaries: Locating the Modern City*, Minneapolis, MN: University of Minnesota Press, 2007.
Clarke, David B., ed. *The Cinematic City*. London: Routledge, 1997.
Coş, Nezih. "Türkiye'de Sinemaların dağılışı." *As Akademik Sinema*, no. 2 (1969): 19–20.
—. "İstanbul'un Sinemaları…" *As Akademik Sinema*, no. 4 (1969): 11–20.
Dönmez-Colin, Gönül. *Turkish Cinema: Identity, Distance and Belonging*. London: Reaktion, 2008.
Ebiri, Bilge. "How Does It Feel to Feel? Recent Turkish Cinema." *Cinema Scope: Expanding the Frame on International Cinema* 23 (2005). Accessed February 28, 2010. http://www.cinema-scope.com/cs23/fea_ebiri_turkish.htm.
Eiland, Howard. "Reception in Distraction." *Boundary* 2, no. 30 (2003): 51–66.
Erdoğan, Nezih. "Narratives of Resistance: National identity and ambivalence in the Turkish melodrama between 1965 and 1975," *Screen* 39, no. 3 (1998): 259–71.
—. "Narratives of Resistance: National identity and ambivalence in the Turkish melodrama between 1965 and 1975." In *Asian Cinemas: A Reader and Guide*, edited by Dimitris Eleftheriotis and Gary Needham. Honolulu, HI: University of Hawaii Press, 2006.
Erdoğan, Nezih, and Deniz Göktürk. "Turkish Cinema." In *Companion Encyclopedia of Middle Eastern and North African Film*, edited by Oliver Leaman, 533–73. London: Routledge, 2001.
Fitzmaurice, Tony. "Film and Urban Societies in a Global Context." In *Cinema and the City: Film and Urban Societies in a Global Context*, edited by Mark Shiel and Tony Fitzmaurice. Oxford; Malden, MA: Blackwell, 2001.
Göktürk, Deniz. "Turkish Delight–German Fright: Migrant Identities in Transnational Cinema." In *Mediated Identities*, edited by Karen Ross, Deniz Derman, and Nevena Dakovic, 131–49. Istanbul: Bilgi University Press, 2001.
—. "Beyond Paternalism: Turkish German Traffic in Cinema." In *The German Cinema Book*, edited by Tim Bergfelder, Erica Carter, and Deniz Göktürk, 248–56. London: BFI, 2002.
Grainge, Paul, ed. *Memory and Popular Film*. Manchester, UK; New York: Manchester University Press, 2003.
Güçhan, Gülseren. *Toplumsal Değişme ve Türk sineması: Kente Göç Eden İnsanın Türk Sinemasında Değişen Profili*. Ankara: İmge Kitabevi, 1992.
"Güvenevler dolandırıcı sahipleri ortadan kayboldu." *Akşam*, June 19, 1960, 1.
"Güvenevler inşaati meselesi imar vekaletince ele alındı." *Akşam*, May 8, 1960, 2.
"Güvenevlerin arazisine Belediye sahip çıkıyor." *Akşam*, May 1, 1960, 2.
"Güvenevlerin yaptığı inşaat yıktırılacak." *Akşam*, April 23, 1960, 1.
Hansen, Miriam. "Early Cinema, Late Cinema: Transformations of the Public Sphere." In *Viewing Positions: Ways of Seeing Film*, edited by Linda Williams, 134–52. New Brunswick, NJ: Rutgers University Press, 1995.
Hollander, Anne. *Moving Pictures*. New York: Alfred A. Knopf, 1989.
"İlgi Çekici bir Türk Filmi: Acı Hayat." *Milliyet*, March 31, 1963.
İpek, Bülent. "Erksan'ın 'Acı Hayat'ı İki Diziye Konu Oldu." *Sabah*, October 27, 2005.
İstanbul Ulaşımında 50. Yıl. Ankara: Karayolları Genel Müd., 1974.
Johnson, Vida T. and Graham Petrie. *The Films of Andrei Tarkovsky: A Visual Fugue*. Bloomington; Indianapolis, IN: Indiana University Press, 1994.
Kalemci, Arzu and Şükrü Özen, "Institutional Change in the Turkish Film Industry (1950–2006): The 'Social Exclusion' Impact of Globalization." *TODAIE's Review of Public Administration* 5, no. 1 (March 2011): 69–120.

Kayserilioğlu, Sertaç. *Dersaadet'ten İstanbul'a Tramvay*. Istanbul: İETT Genel Müdürlüğü, 1998.

Keyder, Çağlar, ed. *Istanbul: Between the Global and the Local*. Lanham, MD: Rowman & Littlefield, 1999.

Kır, Semra. *İstanbul'un 100 Filmi*. İstanbul Büyükşehir Belediyesi Kültür A.Ş. Yayınları, 2000.

Kırel, Serpil. *Yeşilçam Öykü Sineması*. Istanbul: Babil, 2005.

Köksal, Özlem, ed. *World Film Locations: Istanbul*. UK: Intellect: 2012.

Kracauer, Siegfried. *Salaried Masses: Duty and Distraction in Weimar Germany*. Translated by Quintin Hoare, introduced by Inka Mülder-Bach. London: Verso, 1998.

Landsberg, Alison. *Prosthetic Memory: The Transformation of American Remembrance in the Age of Mass Culture*. New York: Columbia University Press, 2004.

McQuire, Scott. *The Media City: Media, Architecture, and Urban Space*. London: Sage, 2008.

Makal, Oğuz. *Sinemada Yedinci Adam: Türk Sinemasında İç ve Dış Göç Olayı*. Izmir: Köprü, Mars Matbaası, 1987.

"Manikürcü kız, Türkan Şoray'ın kaderini çizdi," *Milliyet*, June 6, 1985.

Mennel, Barbara. *Cities and Cinema*. Abingdon, Oxford: Routledge, 2008.

Misek, Richard. *Chromatic Cinema: A History of Screen Color*. Chichester, UK: Wiley-Blackwell, 2010.

Özgüç, Agah. *Türk Sinemasında İstanbul*. Istanbul: Horizon International, 2010.

Öztürk, Mehmet. *Sine-Masal Kentler: Sinematografik Bir Üretim Alanı Olarak Kent Üzerine Bir İnceleme*. Istanbul: Om, 2002.

Pamuk, Orhan. *Istanbul: Memories of a City*, translated by Maureen Freely. London: Faber and Faber, 2005.

—. Foreword to *Ara Güler's Istanbul*. London: Thames and Hudson, 2009.

Penz, François, and Maureen Thomas, eds. *Cinema & Architecture: Méliès, Mallet-Stevens, Multimedia*. London: British Film Institute, 1997.

Pérouse, Jean-François, and A. Didem Danış. "Zenginliğin Mekânda Yeni Yansımaları: İstanbul'da Güvenlikli Siteler." *Toplum ve Bilim*, no. 104 (2005): 92–123.

Pursun, Doğan. "Göç konusu Beyaz Perdeye Aktarıldı." *Akşam*, January 5, 1964.

Safa, Peyami. *Fatih-Harbiye*. İstanbul: Ötüken Nesriyat, 1978.

Schwitzgebel, Eric. "Why Did We Think We Dreamed in Black and White?" *Studies in History and Philosophy of Science*, no. 33 (2002): 649–60.

Scognamillo, Giovanni. *Türk Sinema Tarihi, 1896–1997*. Istanbul: Kabalcı, 1998.

Shiel, Mark and Tony Fitzmaurice, eds. *Cinema and the City: Film and Urban Societies in a Global Context, Studies in Urban and Social Change*. Oxford; Malden, MA: Blackwell, 2001.

Simmel, Georg. "The Metropolis and Mental Life (1903)." In *Simmel on Culture: Selected Writings*, edited by David Frisby and Mike Featherstone, 174–85. London; Thousand Oaks, CA: Sage, 1997.

Simpson, Catherine. "Turkish Cinema's Resurgence: The 'Deep Nation' Unravels," *Senses of Cinema* 39 (May 2006). Accessed April 12, 2010. http://sensesofcinema.com/2006/feature-articles/turkish_cinema/.

"Sınıf Atlama Üzerine bir Melodram." *Milliyet*, August 12, 2001.

Suner, Asuman. *Haunted House: Belonging, Identity, and Memory in New Turkish Cinema*. New York: I.B. Tauris, 2010.

Tanyeli, Uğur. *İstanbul 1900-2000: Konutu ve Modernleşmeyi Metropolden Okumak*. İstanbul: Akın Nalça, 2004.

"Taşradan Büyük Şehre Göç: Bir Rejisor İstanbul'a Akın Meselesini Yeni Filmine Konu Olarak Ele Aldı." *Milliyet*, December 23, 1963.

Türeli, Ipek. "Istanbul through Migrants' Eyes." In *Orienting Istanbul: Cultural Capital of Europe?*, edited by Deniz Göktürk, Levent Sosyal, and Ipek Türeli, 144–64. New York; London: Routledge, 2010.

—. "Istanbul in Black and White: Cinematic Memory." In "Cinematic Urbanism." Online component to *Ars Orientalis*, no. 42 (October 2012). Accessed November 1, 2012. http://www.asia.si.edu/research/articles/istanbul-in-black-and-white.asp

—. "Beyoğlu/Pera on a Cinematic Map of Istanbul." In *World Film Locations, Istanbul*, edited by Özlem Köksal, 90–91. London; Chicago, IL: Intellect Books and Chicago University Press, 2012.

Türkali, Vedat. *Eski Filmler*. Istanbul: Gendaş Kültür, 2003.

Türkoğlu, Nurçay, Mehmet Öztürk and Göksel Aymaz, eds. *Kentte Sinema, Sinemada Kent*. Istanbul: Yeni Hayat Yayıncılık, 2004.

Ülgen, Hasan Bedrettin. "Güvenevler mahallesinde halk dert yanıyor." *Akşam*, April 24 1960.

Wharton, Annabel Jane. *Building the Cold War: Hilton International Hotels and Modern Architecture*. Chicago, IL: University of Chicago Press, 2001.

Yanardağoğlu, Eylem. "TV Series and the City: Istanbul as a Market for Local Dreams and Transnational Fantasies." In *Whose City Is That? Culture, Design, Spectacle and Capital in Istanbul*, edited by Dilek Özhan Koçak and Orhan Kemal, 47–63. Newcastle: Cambridge Scholars Publishing, 2014.

4 Exhibiting vernacular heritage

Urban vernacular houses in Turkey were heritagized in the 1970s, a decade in which concrete-frame apartment buildings came to dominate the fabric of cities, especially that of Istanbul, the largest and fastest growing city in the country. Through architectural gallery exhibitions in the 1970s, and via more mainstream media representations (films, TV shows) later on, the traditional wooden house was framed as a cultural asset and shared inheritance that needed to be protected from rampant urbanization.

Take, for example, the Eurimages-backed, memorable feature film *Steam: The Turkish Bath* (*Hamam*, 1997, dir. Ferzan Özpetek). This film makes local discourse on Turkish vernacular architecture available to transnational audiences while marketing a particular exotic image of Istanbul. Its title and publicity images refer to homosexual love in a steamy bath, evoking tropes of Orientalism. But beyond this, the film addresses the anxiety stemming from the loss of old houses and their replacement by speculative real estate development and proposes as a remedy the potential of transnational mobility and collaboration within Europe.

The house is inherited by a young Italian architect, Francesco (played by Alessandro Gassman), from an aunt who, in the footsteps of Levantines, had made the city her home. Indifferent at first, Francesco ends up saving not only the house but also the whole neighborhood from developers who want to replace its fabric of old houses with a characterless modern complex. The film opens with the death of Francesco's aunt. News of her passing is announced through the neighborhood by word of mouth. On screen, the camera follows various conversations through cuts and then rises above the street in a tracking shot to frame it against the city skyline, associating this house and this neighborhood firmly with Istanbul's image. Viewers see that this is a traditional neighborhood of old wooden houses, where people communicate not by mail or cell phone but through personal conversation and by shouting across the street. The film also taps immediately into local imaginings of the Turkish house and the traditional neighborhoods it constitutes.

In contrast, midway through the film, Francesco meets the prospective redeveloper in a high-rise overlooking a modern part of the city. In the meeting room sits an architectural model of the speculative development project that will replace the neighborhood. Upon seeing the model, Francesco is disturbed and declines to sign the contract to sell his inherited property. The camera then cuts to a conversation between neighborhood women similar to that of the pre-title sequence, but in this scene, the households organize and collectively decide against selling their houses. Through such devices, the film implies that resistance to the forces of globalization is born with and through the architecture of the traditional wooden house—enabled by its windows, shutters, balconies, and intimacy of scale.

The film establishes the qualities of such houses (the local) against those of nondescript high-rises (the global) inhabited by sleek but ultimately corrupt developers. It positions the European expert architect, in the person of Francesco, as an intermediary. The city's old wooden houses nurture family, sustain community, enable resistance, and offer a sense of place and renewed selfhood to the European. As "inheritance," the old house and, by extension, the neighborhood and the city are defended by the expert for the locals who live in it against other locals who want to modernize it. As such, the plotline is an inadvertent parable for the process through which Istanbul's old houses came to be regarded as heritage through the efforts and influences of Europe-centered transnational institutions and policy frameworks.

In this chapter, I discuss early efforts to heritagize the traditional wooden houses in the city. Many studies and monographs have contributed to the definition of the house as typology, dubbed variably as the "Ottoman" or "Turkish house"; however, the process by which it became a heritage object and even a "theme" has not received due critical scrutiny. How and why was a type of domestic architecture in Istanbul heritagized? Who were the promoters of this type of heritage? How did they mediate evolving European norms in local and professional contexts, and how did they translate local concerns to international platforms? I identify the local actors, associations, and networks key to the designation of this type as cultural heritage in the 1970s, during the same period in which heritage promotion policies appeared under the influence of Europe-based supranational organizations. My discussion is limited to Istanbul-based actors because dramatic transformations in Istanbul have been the source of discussions on the typology, in fact from earlier on. I use archival material and biographic accounts to trace the interactions of the local actors. I also rely on newspaper and journal coverage—I regard such popular accounts as constitutive of the public debate on the fate of historic homes and, therefore, of the historical account.

The heritagization of vernacular houses in Turkey is connected to post–World War II European preservation discourses, responding to growing concerns about the destruction of historic city centers through rapid urbanization; in the context of Istanbul and Turkey specifically, this also responds to the rise of "civil society," policy decisions, and legal arrangements, all encouraged by a host of relatively new supranational Europe-based organizations in the 1970s.[1] Heritagization, here, refers to heritage not merely as an object or place but as a social construct and process.[2] These local actors, to whom I refer to as "Istanbul enthusiasts" (*tutkunları*), were encouraged by historic preservation projects they observed abroad in Europe and motivated in person by representatives of supranational organizations. They would adopt and distort the simultaneously developing international norms, which shifted from individual monument-based restoration to area conservation in the postwar era. Supranational institutions supported historic preservation with the aim of fostering the notion of a common European and "universal" inheritance; their efforts in "democratizing" heritage found popular support in other European countries, but local actors in Istanbul cast their objectives in nationalist, and sometimes elitist, tones—almost never discussing the multicultural, multi-religious makeup of the city before non-Muslim minority communities were gradually forced out. In addition, while international norms and legislations of the 1970s were adopted swiftly in Turkey, they were not necessarily implemented. Inadvertently, these local actors' early efforts paved the way for the government-led gentrification of certain streets and neighborhoods, as well as providing inspiration for *mahalle*, or neighborhood, TV series and films and source imagery for commodification in the real estate market in the following decades.

Heritage and inheritance

Heritage, once the privilege of the rich who would "inherit" it from ancestors, has been democratized in modern times.[3] The modern conception of heritage is expedient: As a shared value, it strengthens bias among citizens for their nation or faith, inviting them to realize, together with other citizens, that they are the inheritors of a particular past and, therefore, share a common future. Another aspect of architectural heritage is that it justifies territorial claims. Artifacts from antiquity, including whole buildings, were first "collected" by European nation-states to construct a genealogical tree, at the apex of which they would place themselves. The Ottoman state would start its own archeological collections in response to European imperialist infringement to which archeological excavations provided a pretext.[4] Hence the first legislation in Turkey dealt with the status of artifacts from antiquity (Antiques Regulation, Asar-ı Atika Nizamnamesi, 1874). The elevation of the urban vernacular buildings, in their urban context, to heritage status came much later.[5] The next major law, of 1973, paralleled developments in Europe in promoting the preservation of buildings, not in isolation but in the context of their area or site.

Just as European and, later, postcolonial nation-states became involved in the making of heritage, the postwar merging Europe too has engaged in heritage production to consolidate continental identification.[6] The union of Europe is defined by values of cultural diversity and multiculturalism, but it also demands a measure of political-economic convergence—its motto is "unity in diversity." Thus, according to the trio of Graham, Ashworth, and Tunbridge, "the need for a European heritage" stems from a discord between integration and the lack of corresponding legitimizing identification.[7]

Preservation of architectural heritage has been a cornerstone of European integration since the movement to merge began, but the constitution of heritage has changed radically over time. In Istanbul, traditional neighborhoods and old houses were razed in the immediate postwar period without qualms as the "West" hailed Turkey as a successful example of modernization. In 1959 Istanbul even received the Council of Europe's "Europe Prize," created in 1955, for its urban renewal operations. As the European Council president at the time announced:

"We all know the courage and determination of Istanbul, the guard of the Straits, the spectacular rebuilding effort it has undertaken without damaging any of its historical treasures that are the living witness to its bright past."[8] When the prize was awarded, urban renewal was seen as a technique to make Istanbul more "European"—an outcome that matched the prize's stated goal of "promoting European unity." Urban renewal policies isolated the city's monuments and opened vistas to them along newly carved-out boulevards. Yet this process also caused the demolition of much of the city's historic housing stock, making thousands of people homeless.[9]

The discontent and anxieties stemming from rapid urbanization and the loss of old houses gave way over the following decades to a vibrant debate on urban culture. Eventually, increased awareness of the preservation and cultural heritage policies advocated by supranational institutions led a small number of professionals and enthusiasts to advocate for a return to the Turkish house and a rediscovery of old neighborhoods (see Figure 4.1). But the coupling of this professional call with market reforms in the post-1980 period soon led to types of nostalgia that, in Svetlana Boym's terms, were more "restorative" than "reflective."[10]

Figure 4.1 Call for action for the protection of endangered historic homes.

Poster for "The Second Turkish Houses Week" organized in 1984 by Türkiye Tarihi Evleri Koruma Derneği (Türkev).

For my discussion here, what is important is that the democratization of heritage has allowed individuals who may not have personal ties to any particular historic home to articulate positions and form new alliances and associations around the fate of old houses and the neighborhoods they constituted. In this sense, the cases from Istanbul that I discuss are specific to the city of Istanbul, and to Turkey at large, but also symptomatic of transnational developments. To give an example close to Turkey, anthropologist Christa Salamandra discusses the revalorization of the "Old Damascus"—that is, the historic core of Damascus—among the Damascene elite in the 1990s.[11] She characterizes "The Society of Friends of Damascus," established in 1977, as an elite group established for socialization more than activism, which is aimed at preservation. Here, appreciation of Old Damascus provided a loose group boundary or "distinction" *à la* Bourdieu despite the fact that the group's members did not live in Old Damascus. In Damascus, elite Damascenes try to distinguish themselves from newcomers from the countryside by way of the importance they attached to "urbanite culture" and architectural heritage as its manifestation. This notion of distinction contradicts the idea of sharing a common future or democratization. Heritage controversies and activism revolve around this contradiction: i.e. whose heritage is erased and/or claimed by whom? An elite group's embracing of architectural heritage at a time of its waning influence is "restorative" in this sense—that is, it seeks to restore the group's relative position. However, such a reading can only partially explain why supranational organizations and national governments have been supportive of architectural heritage preservation initiatives.

The institutionalization of historic preservation since the 1970s has corresponded with the turn to tourism in deindustrializing cities as well as urban renewal schemes that seek to revitalize inner city areas. Historic preservation of the urban vernacular can be regarded as the leading arena of "culture"—a category used to market cities, part and parcel of the global turn from managerial to entrepreneurial city governance. Historic preservation-led urban renewal in inner-city neighborhoods pushes out poorer residents and helps sanitize them in predictable ways amenable for international tourists, as well as rendering them welcoming to affluent residents. While much has been written about historic preservation-led urban renewal and state-led gentrification in Istanbul since the 1990s, how experts and Istanbul-enthusiasts in Turkey led the way in the protection of historic homes and created area conservation models is a lesser-known account.

Out of the several cases and figures I discuss, two—Oya Kılıç's exhibition "Istanbul 1800" in 1975 and Perihan Balcı's multiple exhibitions on "Istanbul's Old Houses and Streets"—are little-known and unique examples of women's contribution to the historic preservation of the urban vernacular in Turkey. Another case I discuss, the Touring and Automobile Club's restoration of Soğukçeşme Street is the most well known, having been featured in international media outlets, and because of the prominent location of the street between Topkapı Palace wall and Hagia Sophia. Even then, the usual narrative follows that of the Club's Chairperson that it was his own vision. I show that in fact supranational organizations' efforts, and the central government's interest in tourism revenue-generating projects, have coalesced with experts' and enthusiasts' calls to protect disappearing architectural expressions of a shared national culture.

"Ottoman/Turkish House" as a vernacular type

The earliest writings on vernacular architecture in the late Ottoman Empire date from the first decades of the twentieth century; these focus on wooden urban houses, the traditional neighborhoods they constituted, and the character of old Istanbul.[12] In a period when most of the middle classes desired to move out of old houses to modern apartment buildings, the topic of wooden houses was embraced by only a handful of intellectuals in the same circle. One of their common objectives in looking at the vernacular was to derive from it a contemporary, modern, and simultaneously national idiom for new housing. These early commentators, including Celal Esad Arseven (starting with *Constantinople*, Paris, H. Laurens, 1909), Hamdullah Suphi (Tanrıöver), Doctor Rıfat Osman, Doctor Süheyl Ünver, Mübarek Galib (Eldem), and Architect Arif (Koyunoğlu), precede Sedad Hakkı Eldem's studies and famous "National Architecture Seminar" of the 1930s at the Fine Arts Academy, where the construct of the Turkish house took center stage. How Eldem and his students documented surviving examples of significant urban vernacular houses, produced typological studies based on this documentation, and derived from these studies his national yet modern idiom to apply in numerous built and unbuilt projects is well documented through the seminal work of architectural historian Sibel Bozdoğan.[13] The documentation and categorization of old wooden houses as Turkish house(s) was connected to a disciplinary development whereby this group of residential buildings was first defined as other to (monumental examples of) architecture to be later assimilated back into the discipline as a source or point of origin.[14]

Today, the house is viewed as a "site of memory" rather than an architectural form or type with any precise definition.[15] In the context of Istanbul, nostalgia for the city's lost character centers on the Turkish house and the traditional *mahalle*, or neighborhood, it constituted.[16] The *mahalle* is imagined to have provided a "sense of place," a sense of belonging and identity.[17] There is a great diversity of vernacular housing forms across Turkey. But because of their powerful role as a signifier of lost social values, it is Istanbul's wooden houses from the eighteenth and nineteenth centuries that most lend their look to the apparition of the Turkish house.

While Eldem referred to Ottoman-era houses as "Turkish" with overt nationalistic connotations, several other scholars, such as Ayda Arel (1982) and Doğan Kuban (1975, 1995), have opted since the mid-1970s for "Ottoman" as the umbrella term, in partial reaction to Balkan countries' efforts to define the houses as part of their national heritage, as in "Bulgarian Houses" or "Greek Houses"; however, the signifier "Turkish" has not been totally abandoned.[18]

According to monographs that describe the house as a "fixed entity," as a "type," it consists of two or more floors.[19] The first floor is walled off from the street and is where public social life and (in semi-rural environments) cooking takes place. The second floor provides the living quarters of the family. It includes a hall called the *hayat* (life, living) or *sofa*, and various other rooms. The entrance to each room features a narrow service area with a built-in cupboard to stow bedding and other household items. Beyond this service area, rooms may have slightly raised rectangular platforms for sitting, surrounded by low, fixed seating. In each room, walls, ceilings, and floors are also subtly partitioned into zones, and overall, their decoration and design emphasize horizontality. Based on these characteristics, the rooms are characterized as self-sufficient units, and the house in general is praised for its minimalism and multi-functionality. However, such descriptions are largely constructions that mainly serve a professional agenda to derive principles for contemporary architecture.

Heritagization of the Ottoman/Turkish House

Heritagization of old wooden houses is connected to the development of expert visions for an open-air museum in the historic peninsula. The wooden vernacular, prevalent in the historic peninsula, had already come to signify Muslim Ottoman cultural milieu by the early twentieth century (as briefly mentioned in Chapter 3)—in contradistinction to newer, modern districts such as Pera and Harbiye to the north of the Golden Horn, which came to be regarded as the "European" part of the city—because, first, the sale of houses by Muslims to non-Muslims was prohibited well into the nineteenth century, and, second, Ottomans decreed the use of wooden construction in residential properties for its relative safety in this earthquake-prone city.[20]

The first expert proposal to conserve parts of the historic peninsula as an open-air (archeological) park was made by the French planners Donald Alfred Agache and Jacques-Henri Lambert as early as 1934 for Sultanahmet Square, formerly the site of the Roman hippodrome.[21] Another French planner, Henri Prost, envisioned in his 1938 master plan for the city the construction there of a Republic Square as the site of modern state ceremonies. Ultimately, his proposal was not accepted, and Taksim Square was preferred for the same function. In a new plan in 1947, Prost then proposed turning Sultanahmet into an "archaeology park," but this proposal too was not implemented.[22] Interestingly, Albert Gabriel took the latter proposal for an open-air archeology park to UNESCO in 1947 on the basis that "the conservation of pre-Turkish structure can only be undertaken with foreign funds" due to lack of local funds.[23] To convince a hesitant UNESCO, the purpose of which was maintaining peace rather than preserving culture, UNESCO appointee La Charriere wrote in 1949 "for UNESCO, this would be an enormous publicity if they can manage to transform this section of Istanbul into a symbol of utmost tolerance and fusion of cultures, the peaceful image of which Jerusalem could never generate."[24] The talks between UNESCO and the Turkish government were abandoned by 1953 due to a number of factors, including inadequate diplomacy.[25] While a failed effort, it nevertheless shows the evolving motivation of a supranational (still, Europe- and Paris-based) organization such as UNESCO in the preservation and display of architectural heritage.

Another expert proposal appeared in the late 1970s under the guidance of the European Council.[26] Its Assembly initiated an international campaign in 1978 to preserve Istanbul's architectural heritage.[27] The campaign identified Topkapı, Sultanahmet, Zeyrek, Süleymaniye, and Yedikule as conservation areas. UNESCO subsequently contributed funds for local experts to study and prepare targeted plans for these neighborhoods.[28] Turkey's Ministry of Tourism contributed to this effort by commissioning proposals from Istanbul Technical University's (ITU) Restoration Program for the rehabilitation of selected streets in Sultanahmet. The proposals included returning the neighborhoods to their nineteenth-century state—with streets paved with cobblestones instead of asphalt, lanterns (*kandil* and *fener*) instead of electric lights, and the elimination of motor vehicles in favor of foot traffic and horse carriages. The proposals also envisaged repairing historic homes and reprogramming them for tourism.[29]

The 1978 vision, as encouragingly reported in the newspapers of the time, bears some similarity to that of the current government's much-contested Museum-City Project, launched in the mid-2000s, and accompanying legislation, which allows local governments to declare historic inner-city neighborhoods as pioneer urban renewal areas.[30] However, the politics of neo-Ottoman neighborhoods are now especially suspect because they come from central and local governments run by the same conservative political party—the AKP. Experts and journalists alike have been pointing out the exclusionary aspects of urban regeneration-based, government-led gentrification and the displacement of existing residents, as well as the ideological limitations of such architectural visions. Of the renewal areas in the historic peninsula, that of the Sulukule neighborhood was widely publicized in Turkey and abroad because the resident Roma population was particularly vulnerable; their displacement meant the dissolution of their unique community and culture.[31]

Referring to the Turkish Monument and Historic Buildings Act of 1973 and other pieces of legislation, Ashworth and Tunbridge remark on "the similarity of timing and content of the key pieces of national legislation" on urban conservation across Europe—in countries "with otherwise distinctly different traditions of planning."[32] One reason for this was that UNESCO and the Council of Europe played key roles in the postwar period in defining expert opinion on architecture and urban planning in many countries, including Turkey, by organizing platforms for the active interchange of ideas. Only recently, however, has UNESCO's cosmopolitan, elitist appropriation of culture elicited scholarly scrutiny with regard to discrepancies between who produces, designates, promotes, and consumes heritage.

Supranational mediators: UNESCO and the Council of Europe

Both UNESCO and the Council of Europe have tried to mobilize state actors to take measures to protect their heritage. Especially for the latter, political integration could not be fully realized without cultural integration, in which architectural heritage plays a central role. These organizations have provided leadership by designating national representatives in different member countries, guiding legislation, encouraging the establishment of local organizations where there were none, and making direct contributions to the preservation of architectural monuments and sites of cultural significance. Underlying their emphasis on culture has also been a recognition that heritage can be a resource for socioeconomic gain.

The conservation of artifacts and buildings has a long history in the "West," but it emerged as a profession only after the Second International Congress of Architects convened in Venice in 1964 (not to omit the importance of the Athens Charter for the Restoration of Historic Monuments of 1931; the Hague Convention for the Protection of Cultural Property in the Event of Armed Conflict in 1954; and of ICCROM, International Center for the Study of the Preservation and Restoration of Cultural Property, created in 1959). The document that followed, known as the Venice Charter, set standards for conservation, mainly broadening the definition of what needed to be preserved from monument (*anıt*) to heritage (*miras*), deemphasizing local and national in favor of the "universal," and extending concern from individual buildings to areas of integrated fabric, including vernacular buildings—notably coinciding with Bernard Rudolfsky's MoMA exhibition "Architecture without Architects," which sought to expand architecture by including vernacular examples from around the world, and signaled the institutional acceptance of vernacular types by design professionals as a critique of international modernism and the negative impact of modernist planning in postwar cities. While the Venice Charter is cited today as an international document, the Congress was dominated by European experts, and the charter reflected as a result continental European concerns such as the "rehabilitation of destroyed cities" and the promotion of cultural tourism as an economic asset.[33]

The institution that was created to oversee work toward the new goals set by the Charter was the International Council of Monuments and Sites (ICOMOS).[34] It was established to advise UNESCO and be "an international non-governmental organization of professionals, dedicated to the conservation of the world's historic monuments and sites."[35] While UNESCO's strategy was to serve as an arbiter and compile its own list of worthy sites, the Council of Europe decided to also enlist local civil-society organizations in the task. Both have now emerged as "authorizing institutions of heritage."[36]

The Council of Europe undertook many important initiatives in the heritage field during the 1970s.[37] Europa Nostra (Our Europe) had been founded in 1963 in the Office of the Council of Europe in Paris by a group of nongovernmental heritage organizations headed by Italia Nostra. It established an awards scheme in 1978 to acknowledge and promote outstanding heritage preservation efforts.[38] But one of the most influential initiatives advanced by the Council of Europe was its designation of the year 1975 as the European Architectural Heritage Year (EAHY).[39] Beginning in 1972, the council spearheaded a publicity campaign that led to pilot preservation projects, publications, and conferences. These culminated in October 1975 with the convening of the Amsterdam Congress under the auspices of the Council of Europe.

The declaration presented at this event (the European Charter of Architectural Heritage) included a set of recommendations to parliaments and institutions across Europe.[40] These introduced the concept of integrated heritage conservation, along with the notion that Europe's architectural heritage was common to all European peoples.[41] Heritage, as redefined in the declaration, was no longer limited to works of outstanding universal value (or monumental expression). The charter also stipulated that respect for and understanding of architectural heritage was necessary "to achieve a greater unity." The charter further acknowledged that this heritage was in danger of disappearing.

Each member of the council was expected to take measures to document and protect its architectural heritage. According to John Delafons, the EAHY "stimulated in England a longer-lasting interest in conservation at the grass roots level. It was perhaps the beginning of the popular concern for conservation which increasingly supplanted the elitist tradition of conservation in Britain."[42] In Turkey, the preparation for the EAHY culminated in the establishment of a governmental-level national committee that recommended Istanbul as the site of one of three pilot heritage-conservation projects.[43] Educational establishments of architecture in Istanbul organized conferences and exhibitions and produced publications to define architectural heritage. The EAHY stimulated, not necessarily a bottom-up movement, but certainly a number of projects, events, and associations that aimed at promotion and publicity.

According to Turkish professionals, the most notable change signaled by the EAHY and the Amsterdam Congress was a shift from simply protecting historic monuments to more integrated efforts to conserve historical environments.[44] In the years that followed, Turkish delegates who attended similar international symposia and exhibitions organized under the auspices of UNESCO and the Council of Europe reported on their experiences and learning in Turkish journals of architecture. These delegates, together with enthusiasts, also played an active role in organizing local events to highlight the architectural heritage of Istanbul. Occasionally they took these exhibitions back to the European Council.[45] International congresses also helped raise awareness among local politicians about the potential of heritage initiatives. Thus, despite the increased politicization of the Chamber of Turkish Architects, which corresponded with the radicalization of politics in Turkey, the pendulum of expert attention began to swing away from ideals of social equity and access toward (the externalization and problematization of) "culture."[46]

As part of this shift, Turkish delegates often reported that Balkan countries were claiming Turkish heritage as their own.[47] At the same time, they were impressed by the conservation efforts in Europe and proposed the expropriation of historic neighborhoods in Turkey to transform them into open-air museums with functional reprogramming. Thus, local actors rearticulated international calls for heritage preservation in nationalist terms and supported revenue-generating proposals rather than those aimed at social sustainability, e.g. that would empower existing residents to take care of their own homes.

An open-air museum

Proposals for an open-air museum to exhibit Old Istanbul were quickly embraced. Tourism was cited as an economic justification in the daily press: One journalist suggested that an open-air museum in the city would in effect act as a "foreign-exchange mint" (*döviz matbaası*) in order to highlight the expediency of heritage preservation in an economy based on an import substitution model and short of foreign exchange.[48] The "Istanbul 1800" exhibition (May 21–June 20, 1975) proposed preserving a part of the city as it looked in the year 1800, "before [it lost] characteristics with the impact of Europeanization."[49] Organized by Oya Kılıç, a young female graduate of a local interior architecture school, and held at the centrally located Galatasaray Branch Gallery of the Yapı Kredi (Building and Credit) Bank, the exhibition first made news in early 1974 and inspired supportive write-ups even after it was over (see Figure 4.2).[50] Kılıç had fond memories of Madurodam, the nation-themed miniature park, she had visited in Holland as a child. It was in a UNESCO newsletter that she first encountered the concept of an open-air museum as a young professional. She described the project, not in economic terms, but through patriotic associations. Other more senior Istanbul

Figure 4.2 "Istanbul 1800" exhibition on the cover of arts and culture weekly *Milliyet Sanat*.

Milliyet Sanat, no. 135 (June 1975).

enthusiasts explained the significance of the exhibition as a corrective to rapid and uncontrolled urbanization.

In support of Kılıç's "Istanbul 1800" exhibition, Touring Club Chairperson Çelik Gülersoy helped pay for photographs, and Sedad Eldem, the aforementioned renowned Turkish architect who devoted his career to the development of a national architecture based on typological studies of old wooden houses, allowed Kılıç access to his extensive private archive. Istanbul enthusiast *par excellence* Reşat Ekrem Koçu wrote supportively even before the exhibition took place: "Istanbul is transforming by the day and getting uglier in the hands of some people blinded with ambition for money and profit … Oya [Kılıç]'s aim is to cherish that vanishing Turkish Istanbul."[51]

Eldem was highly appreciative of Kılıç's proposal and wishfully predicted, "it is women who will protect our works of art and houses."[52] The role of women in heritage activism is better documented in North America.[53] In the Turkish context only a few professionals are credited, such as Cahide Tamer, Selma Emler, and Mualla Eyüboğlu Anhegger—the trio were among the first female architects in Turkey, and they collaborated in the restoration of Rumeli Hisar. Emler represented Turkey in the 1964 Venice meeting, along with architectural historian Doğan Kuban.[54] These leading restoration architects' work consisted mostly of monumental buildings. Kılıç's and Perihan Balcı's efforts differ, first by targeting the preservation of vernacular houses, and second by engaging promotion and instigating public debate through exhibitions and publications rather than direct professional intervention.

"Istanbul 1800" included black-and-white photographs, sketches by Kılıç, measured drawings, two architectural models—one in 1/500 scale of the area around Süleymaniye, from Şehzadebaşı to Vefa, and another smaller one of a stereotypical segment of the Bosphorus with its waterfront mansions, *yalıs*—and what she called "authentic artifacts" such as doorknobs, window grills, and wall carvings. Kılıç explained:

> My aim is to create a historic and touristic open-air museum and cultural center that I named "Istanbul 1800." According to the information given by ICOM (International Council of Museums) that is under UNESCO, there are 152 open-air museums in 14 countries only on the European continent. In all of them, the goal is to exhibit a culture that is disappearing.[55]

Appealing to and simultaneously reflecting popular nationalist sentiments, perhaps following the celebration of 1973 as the half-century mark since the establishment of the Republic, she argued that the protection of monuments, as opposed to vernacular architecture, undermined Turkey's claim to the city. She suggested that the old fabric of the historic city was created during the five hundred years of Ottoman rule, but that its obliteration had been a result both of modernization and new development and the emphasis on the preservation of its Byzantine walls and monuments. She went on to contend that the historic Turkish houses were prime "examples that would prove that Istanbul [was] ours for the past five hundred years." In the same breath, she referred to old Turkish houses as "cultural weapons." Their touristic potential was frequently invoked in write-ups that urged the government to take action on the open-air museum.

This young architect's call was reflective of an aspiration for cultural expression that emerged out of a new type of European identification. Her argument revealed her familiarity with contemporary debates on museology and heritage conservation. The folk museums she had in mind, those from Nordic countries, were responses to industrial modernity that sought to protect rural architecture. Even recent examples of such a form of historic conservation, as in the case of pilot cities in Italy, tended to focus on provincial towns without much economic importance. In Turkey, however, Kılıç's exhibition proposed conserving living districts in the heart of a fast-growing city. Ironically, the exhibition was hosted by a bank that was heavily invested in the process of building and selling new residential settlements outside the historical core. Customer-visitors to the exhibition in the bank's gallery in Beyoğlu, in the "European" northern part of town, could easily imagined a touristic open-air museum in the historic core, since they had already abandoned it or planned to do so for homes elsewhere.

"Civil society" for architectural heritage

As already mentioned, the Council of Europe encouraged a bottom-up approach to conservation. But most of the members of the local organizations concerned with the protection of old houses of the city were academics and professionals who did not live in the houses they sought to protect. They were, rather, associations of enthusiasts, more concerned with appearances than with the actual lives lived within these houses.

Until the mid-1970s, isolated monuments had occupied the focus of conservation efforts in the city. These were overseen by the High Council of Monuments and Sites (1952, Anıtlar Yüksek Kurumu), which was connected to the Ministry of Education. However, with the popularization of the notion of architectural heritage, various individuals, groups, and institutions began taking a more active, and sometimes a leading, role in conservation efforts. Among these, the Touring Club's intervention in urban environments started with the rehabilitation of the houses surrounding the Byzantine church-turned-mosque of Saint Saviour in Chora (Kariye). It also became nationally and internationally famous for its renovation of Soğukçeşme Street. But two other nongovernmental organizations were founded in 1976: The Foundation for Monuments, Environment, and Tourism (Türkiye Anıt Çevre Turizm Değerleri Vakfı, TAÇ); and the Association for the Protection of Historical Homes (Türkiye Tarihi Evleri Koruma Derneği, Türkev).[56] Both of these organizations incorporated images of the Turkish house into their logos (see Figure 4.3), and their members consisted largely of academics within the field of architecture who were disturbed by profit-driven urban expansion and simultaneously inspired by preservation efforts in Europe, especially in neighboring Balkan countries. Encounters in Europe or with European institutions were the primary influences on the founding of organizations such as TAÇ and Türkev, as opposed to anyone's personal attachment to an inherited house.

Figure 4.3 Türkev logo featuring a stylized wooden house.

Cover of a booklet by Türkev with logo stylizing the wooden house at the center.

The case of Türkev is especially revealing. Although not an Istanbulite by birth, its founder, Perihan Balcı, grew up in Istanbul and partook in its warm *mahalle* sociability. After a personal loss in 1965, she began taking photographs of the city's disappearing historic houses. This attempt to document and exhibit the disappearing city—a therapeutic effort, indeed—served to delay the imminent end of the old houses. In 1975, Balcı published a photography book titled *Old Istanbul Houses and Bosphorus Yalıs.*[57] In the preface, architectural historian

Doğan Kuban, also an advocate of preservation and a founding member of Türkev, describes Balcı's photography as an antidote to speculative urbanism and a means of educating the wider public. Balcı's work then led to an exhibition at the Fine Arts Academy as part of the Academy's EAHY activities. This led to another exhibition in Ankara, and another in France.[58] With the encouragement of Europa Nostra representatives, in 1976 Balcı then established an association in Istanbul to inform public opinion on the preservation of old houses.

What began as an individual effort to document the vanishing wooden houses of Istanbul transformed into an association promoting the "Turkish" house. Türkev did not involve a bottom-up, grassroots response to the protection of historic housing. Aside from Balcı, all its founding members were prominent professors—all of them male, two of them medical doctors, and the other two architecture faculty. The association maintained a similar contributor profile in the years that followed.[59] The association had three goals: to protect historic homes; to help their owners with legal and financial issues; and to persuade the public to once again take residence in Turkish houses. Türkev eventually became known for such activities as Balcı's exhibitions and lectures displaying Turkish houses at home and abroad; annual "Historic Turkish Houses Weeks," which started in 1983 (see Figure 4.4); informational tours for architecture students organized through collaborations with universities; and the restoration of the Ismail Dede Efendi House, a house museum dedicated to the eighteenth-century Ottoman musician. A "housewife" by self-definition, Balcı was accepted, welcomed, and supported by academics and professionals.[60] She led the association until the mid-1990s, when Cengiz Eruzun, a professor of restoration at Mimar Sinan University (formerly Fine Arts Academy where Eldem studied and later taught) took over. Eruzun then became the project leader of the Municipality's "Museum-City" project in 2006.[61]

TAÇ's former chairperson, Sinan Genim, also a professor of restoration at Mimar Sinan University, attributes the founding of his organization to an international academic meeting in Budapest in 1975. As recollected by Genim, a group of approximately thirty academics squeezed in sightseeing while travelling together by bus to a conference in Budapest. In many Yugoslavian and Bulgarian towns, such as Filibe (Plovdiv), the academics encountered well-preserved examples of Ottoman-Turkish vernacular architecture.[62] These sights and experiences led Genim and his colleagues to establish TAÇ in 1976, with financial help, guidance, and a task list from the Ministry of Tourism—this task list included a conservation study of Soğukçeşme Street.[63] TAÇ has operated in several areas. These have included the production of publications such as Sedad Hakkı Eldem's comprehensive three-volume *Turkish House* (1984–7); the opening of a documentation center to study historic sites; the organization of an annual design competition to encourage the study of historic environments in architectural education; and the restoration of a limited number of mansions.

A model Ottoman street

Soon after the founding of TAÇ, the Ministry of Tourism approached Istanbul Technical University to develop a proposal for the rehabilitation of Soğukçeşme

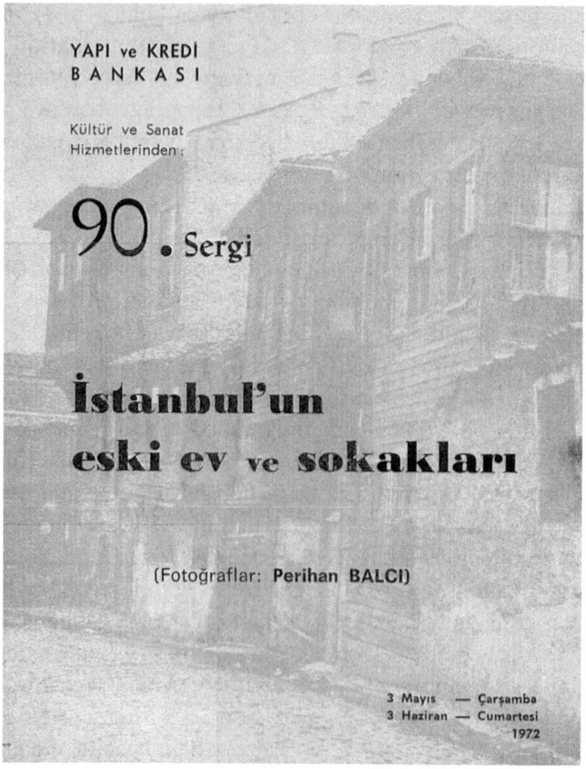

Figure 4.4 Brochure for Perihan Balcı's exhibition "Old Houses and Streets of Istanbul." "Istanbul'un Eski Ev ve Sokakları," held at Yapı ve Kredi Bankası kültür ve sanat hizmetlerinden 90. sergi, May 3–June 3, 1972.

Street and its surrounding neighborhood. The invitation came in response to calls for the development of an open-air museum to showcase the historic fabric of the city for touristic consumption. The result of the ITU study, published in 1979, was the suggestion that tourist-oriented facilities be built along Soğukçeşme Street and that its existing structures be rebuilt with additions to make the street appear more authentic.[64]

Despite these fragmented beginnings, the Soğukçeşme Street project was then adopted and funded by the Touring Club and became the club's most publicized and popular project. Because it also set a precedent for historic preservation for touristic purposes in Turkey, it is important to understand how it came about and how it was received. As its full name makes clear, the Club was an organization established in 1923 to cross-promote tourism and the automobile sector in the then-nascent Republic of Turkey. From 1971 until 1990, the organization was mostly financed with fees charged to Turkish guest workers in Europe when they returned to Turkey for vacation. This income was translated to investments to promote tourism under Chairperson Çelik Gülersoy.

Architects had a love-hate relationship with Gülersoy, another Old Istanbul enthusiast. Despite being educated as a lawyer, Gülersoy identified himself, and was in turn identified as, an urban historian, an art historian, an architect, a planner, a specialist of restoration, an Istanbul gentlemen, and so forth. Among other activities, he authored a series of books on Istanbul and established a library on Soğukçeşme Street dedicated to the study of Istanbul. Using the resources of the Club, he also acted as a model public persona on issues concerning Istanbul. Gülersoy's first major architectural project was the Malta Pavilion, in the Yıldız Palace Park. This received the Council of Europe's Europa Nostra award and validated his authority on matters of historic preservation. Just before the military coup of 1980, after which Turkey shifted from a state-dominated to a privatized economy, the Club acquired a series of important historic preservation contracts that gave Gülersoy more public visibility. He would lease and maintain historic properties and open them to the general public. While these efforts were generally appreciated, architects criticized the mode of "restoration" he chose for Soğukçeşme Street. His acerbic rebuttals and inability to take criticism provided for animated discussions in the architectural media.[65] Paradoxically, it was Soğukçeşme Street that brought national and international acclaim to both the organization and Chairman Gülersoy. The *New York Times* called him "a latter-day Prospero."[66] He single-handedly transformed the Club into a nostalgia machine for Old Istanbul until the 1990s.[67]

The Touring Club bought the properties on Soğukçeşme Street precisely because of the location's touristic potential. The Club then proceeded to demolish the properties and build them anew in concrete frame with only semblance to the street's nineteenth century depictions. In her critical analysis of this mode of restoration, architectural historian Zeynep Çelik situates Soğukçeşme Street within the legacy of nineteenth-century world expositions, and she compares it to the Oriental streets reconstructed for those occasions.[68] The streetscape looked like a row of traditional houses clad in timber and painted in soft pastel colors, but inside, they were commercial buildings with interiors showcasing fashionably ornate 1980s' furniture. The project was promoted as a response to the perceived loss of Istanbul's character—as one of many celebratory news reports presented on the front page of *Cumhuriyet*, Soğukçeşme was "the only street in Istanbul without apartment buildings" (see Figure 4.5).[69] It was realized with the approval of the government during a time when the municipality was simultaneously engaged in a drastic program of demolitions within the historic fabric in an effort to turn Istanbul into a regional center.

The Soğukçeşme Street renovation project was initiated in response to rapid urban transformation and inspired by the European integration process (see Figure 4.6). Once implemented, it became an important site for cultural and touristic consumption, and it paved the way for a wider practice of commodified nostalgia. Over the past decades this has evolved from proposals for the fictional restoration of urban fragments to the wholesale freezing of the historic city. The timing of the reconstruction aligned with the adoption of "neo-Ottomanism" as state policy during the government of Turgut Özal (prime minister,

İstanbul'da apartman olmayan tek sokak

3. Selim döneminden kalma çeşmeden adını alan sokaktaki evlerden 9'u, pastel renklere boyanarak, 120 yataklı bir pansiyona dönüştürüldü. Ayrıca bir İstanbul Kitaplığı kuruldu, Bizans Sarnıcı ise taverna olarak yeniden düzenlendi.

Turing tarafından restore edilen evlerin bulunduğu sokağın açılışında konuşan Çelik Gülersoy, sokağı, "İçinde apartman bulunmayan tek İstanbul sokağı" olarak niteledi. Sokağın restorasyonuna karşı olduğu bilinen Belediye Başkanı Bedrettin Dalan, İsviçre'ye gittiği için açılışa katılmadı.

Haber Merkezi — Türkiye Turing ve Otomobil Kurumu'nun Ayasofya Müzesi'nin arkasındaki Soğukçeşme Sokağı'nda satın alıp onardığı eski İstanbul evleri, turistik pansiyon olarak dün hizme-

(Arkası Sa. 12, Sü. 4'te)

Figure 4.5 "The only street in Istanbul without apartment buildings." A front page newspaper report on Soğukçeşme Street.

Cumhuriyet, October 15, 1986, 1.

1983–9; president, 1989–93), a conservative, nationalist leader and contemporary of Margaret Thatcher in the United Kingdom and Ronald Reagan in the United States.[70] Policymakers like and around Özal emphasized Turkey's Ottoman legacy and its Muslim character, both to counter rising internal ethnic conflict and shape Turkish foreign policy. Hence, the Istanbul Municipality's concurrent urban renewal operations, which involved massive demolition and displacement, especially around the Golden Horn, ironically went hand in hand with historic preservation efforts by experts and enthusiasts. Clearing operations would make Istanbul fit to be a regional center, and preservation would highlight its uniqueness as a city.

The expediency of exhibitions

The Ottoman/Turkish house was a category first constructed within the nation-building policies of the late Ottoman and early Republican eras—that is, in the first decades of the twentieth century. It captured the imagination of a small number of highly influential intellectuals and architects based in Istanbul, at a time when most of the population of the city desired to move to modern apartments. The house gained a special status in melancholic depictions of modern Istanbul. There were suggestions for the preservation of important examples of historic

Figure 4.6 Soğukçeşme Street after restoration on the cover of the official publication of the Touring Club (1987).

wooden houses; however, the notion of deriving from the architectural "type" a revivalist contemporary architecture remained the prevalent motivation.

By the 1970s, growing concerns about the destruction of historic centers through rapid urbanization led in Istanbul, as in many cities in Europe and elsewhere, to the formal re-construction of the house as a heritage object. Exhibitions of the city's old houses and streets were crucial to the heritagization of the wooden vernacular. At a personal level, for enthusiasts, documenting disappearing old houses served to delay the imminent end of *mahalle* sociability. Collectively, for those organizing in associations around their affection for and attachment to "vanishing Istanbul," the appreciation of the aesthetics of wooden houses became a socially distinguishing signifier. Calls by enthusiasts and professionals to protect these houses in their urban context—i.e. through area conservation—were embraced by the government for its potential to generate tourism income and inner-city revitalization.

I have discussed several key local actors, the associations they set up, and their networks in the 1970s to show how evolving European norms were reinterpreted as they were translated to the local context. When compared to their counterparts in the early decades of the twentieth century, these actors had restrained from professional agendas; hence my preference to call them enthusiasts. The 1975 EAHY was particularly instrumental in the founding of Türkev and TAÇ in 1976 and the turn of the Touring Club to historic preservation. The document that resulted from the EAHY meeting clearly stated, "The European Architectural heritage is the common property of our continent." Yet, all these Istanbul-based actors and their associations emphasized the national character of their heritage object, the house, and the nationalist imperative to preserve it. The popular preservation movement in Europe was at least in part to be regarded as an expression for more social equity and the democratization of heritage, even if there were counter tendencies, but in Turkey local calls remained paternalistic and somewhat elitist.

Appreciation of and concerns over the disappearance of the city's old houses and streets led to the heritagization of the Ottoman/Turkish house; this process doubled as a critique of profit-driven real estate development. Ironically, restoration work undertaken, as in Soğukçeşme Street, ended up displacing former residents. Overall, these actors' efforts had tremendous impact for decades to come as the "house" proved vital for imagining Old Istanbul, even if it was identified as the Ottoman or Turkish house, and this imagining continues to inform how the city is represented in visual culture and popular media productions, as well as in public debates on urban regeneration in the present day.

Notes

1 For an overview of postwar preservation discourses, see Mrinalini Rajagopalan, "Preservation and Modernity: Competing Perspectives: Contested Histories and the Question of Authenticity," in *The Handbook of Architectural Theory*, eds. Greig Crysler, Stephen Cairns, and Hilde Heynen (London: Sage, 2012), 208–324.
2 On heritage as social construct, see: Barbara Kirshenblatt-Gimblett, *Destination Culture: Tourism, Museums, and Heritage* (Berkeley, CA: University of California Press, 1998); and "World Heritage and Cultural Economics," in *Museum Frictions: Public Cultures/ Global Transformations*, eds. Ivan Karp, Corinne A. Kratz, Lynn Szwaja, and Tomas Ybarra-Frausto (Durham, NC: Duke University Press, 2006), 161–201. On heritage as process, see: Laurajane Smith, *The Uses of Heritage* (London; New York: Routledge, 2006).
3 David Lowenthal, *The Heritage Crusade and the Spoils of History*. Cambridge: Cambridge University Press, 2003.
4 Wendy M. K. Shaw, *Possessors and Possessed: Museums, Archeology, and the Visualization of History in the Late Ottoman Empire* (Berkeley, CA: University of California Press, 2003), 106–107.

5 The first antiques regulation placed all under the control of the Ministry of Education. This legislation was revised in 1906 to address the shortcomings of the 1874 one in covering works from the Turkish-Islamic period including old houses. The 1906 law was adopted by the new republic and remained in effect until 1973. The Directory of Antiques and Museums, established in 1920, was responsible for archeological findings. In 1951 the "Supreme Council for Preservation" (Gayrimenkul Eski Eserler ve Anıtlar Yüksek Kurulu) was established to work in the field of preservation; the main work of this council was inventorying and registering historically significant buildings and restoring monumental ones, usually in isolation from their urban context.

6 David Lowenthal, "Identity, Heritage, and History," in *Commemorations: The Politics of National Identity*, ed. John R. Gillis (Princeton, NJ: Princeton University Press, 1994), 41–57.

7 Brian Graham, Gregory Ashworth, and John Tunbridge, *A Geography of Heritage: Power, Culture and Economy* (London: Oxford University Press, 2000).

8 "İstanbul Yılın Şehri," *Hayat*, no. 7, February 12, 1960, 1. Short news report on the selection of Istanbul as the City of the Year (1959) at the meeting of the Council of Europe on 25 January 1960 in Strasburg. My translation.

9 On the context of this award, Menderes-era urban renewal, see: Burak Boysan, "Politik Hummanın Silinmeyen İzleri: Halkla İlişkiler Stratejisi Olarak İstanbul'un İmarı," *İstanbul*, no. 4 (1993): 84–98.

10 Svetlana Boym, *The Future of Nostalgia* (New York: Basic, 2001).

11 Christa Salamandra, *A New Old Damascus: Authenticity and Distinction in Urban Syria* (Bloomington, IN: Indiana University Press, 2004).

12 Yavuz Sezer, "The Perception of Traditional Ottoman Domestic Architecture as a Category of Historic Heritage and a Source of Inspiration for Architectural Practice (1909–1931)," MA in History of Art and Architecture (Istanbul: Boğaziçi University, 2005).

13 Sibel Bozdoğan, "The Turkish House Reappraised," in *Sedad Eldem: Architect in Turkey, ed. Suha Özkan and Sibel Bozdoğan* (Singapore; New York: Concept Media/Aperture, 1987), 44–75; Sibel Bozdoğan, "Vernacular Architecture and Identity Politics: The Case of the 'Turkish House'," *Traditional Dwellings and Settlements Review* 7, no. 2 (1996): 7–18; Sibel Bozdoğan, "Nationalizing the Modern, Appropriating Vernacular Traditions," in *Modernism and Nation Building: Turkish Architectural Culture in the Early Republic* (Seattle; London: University of Washington Press, 2001), 255–71. See also: Uğur Tanyeli and Bülent Tanju, ed., *Sedad Hakkı Eldem I, II* (Istanbul: Osmanlı Bankası Arşiv ve Araştırma Merkezi, 2008, 2009).

14 Uğur Tanyeli, "Türkiye'de Modernleşme ve Vernaküler Mimari Gelenek: Bir Cumhuriyet Dönemi İkilemi," in *Bilanço 1923–1998: Türkiye Cumhuriyeti'nin 75 Yılına Toplu Bakış Uluslararası Kongresi 1,* ed. Zeynep Rona (Istanbul: Tarih Vakfı Yayınları, 1999), 283–90. For a broader account of the vernacular, see: Robert Brown and Daniel Maudlin, "Concepts of Vernacular Architecture," in *The Handbook of Architectural Theory*, eds. Greig Crysler, Stephen Cairns, and Hilde Heynen (London: Sage Publications, 2011), 340–55.

15 Carel Bertram, *Imagining the Turkish House: Collective Visions of Home* (Austin: University of Texas Press, 2008); Bozdoğan (1987, 1996, 2001); Gülsüm Baydar, "Between Civilization and Culture: Appropriation of Traditional Dwelling Forms in Early Republican Turkey," *Journal of Architectural Education* 47, no. 2 (1993): 66–74; Esra Akcan, *Çeviride Modern Olan* (Istanbul: YKY, 2008), and *Architecture in Translation: Germany, Turkey, and the Modern House* (Durham, NC: Duke University Press, 2012).

16 Amy Mills, *Streets of Memory: Landscape, Tolerance, and National Identity in Istanbul* (Athens, GA: University of Georgia Press, 2010).

17 Ibid.

18 Hülya Yürekli and Ferhan Yürekli, *Türk Evi: Gözlemler-Yorumlar = The Turkish House: A Concise Re-evaluation* (Istanbul: Yapı-Endüstri Merkezi, 2005), 10.

19 Some of the standard local sources on the wooden vernacular house as an architectural type include: *Sedad Hakkı Eldem, Türk Evi Plan Tipleri* (Istanbul: Istanbul Technical University, 1954); Ayda Arel, *Osmanlı Konut Geleneğinde Tarihsel Sorunlar* (Izmir: Ege Üniversitesi Güzel Sanatlar Fakültesi Yayınları, 1982); *Sedad Hakkı Eldem, Türk Evi: Osmanlı Dönemi, Turkish Houses Ottoman Period I, II, III* (Istanbul: Türkiye Anıt, Çevre, Turizm Değerlerini Koruma Vakfı, 1984, 1986, 1987); Önder Küçükerman, *Kendi Mekanının Arayışı İçinde Türk Evi* (Istanbul: Türkiye Turing ve Otomobil Kurumu, 1985); Doğan Kuban, *The Turkish Hayat House* (Istanbul: Muhittin Salih Eren, 1995); Cengiz Bektaş, *Türk Evi* (Istanbul: Yapı Kredi Yayınları, 1996); Reha Günay, *Tradition of the Turkish House and Safranbolu Houses*, trans. Çelen Birkan (Istanbul: Yapı-Endüstri Merkezi, 1998); Yürekli and Yürekli (2005).

20 Reşat Ekrem Koçu, "Ev," in *İstanbul Ansiklopedisi 10* (Istanbul: Koçu Yayınları, 1971), 5400–406.

21 Pierre Pinon, "Henri Prost, Albert Gabriel, Istanbul Archeological Park and the Hippodrome," in *Hippodrome / Atmeydanı, A Stage for Istanbul's History*, eds. Brigitte Pitarakis and Ekrem Işın (Istanbul: Pera Museum, 2010), 152; Murat Gül, *The Emergence of Modern Istanbul: Transformation and Modernisation of a City* (London; New York: I.B. Tauris Academic Studies, 2009), 94.

22 Nur Altınyıldız, "The Architectural Heritage of Istanbul and the Ideology of Preservation," *Muqarnas* 24, no. 1, History and Ideology: Architectural Heritage of the "Lands of Rum" (2007): 292.

23 Quoted in Pinon, "Henri Prost," 154.

24 Ibid., 155.

25 Pinon, "Henri Prost"; Altınyıldız, "The Architectural Heritage of Istanbul."

26 Earlier conservation report(s) were prepared by Doğan Kuban in the 1960s, according to literature on conservation, but these did not have an echo in popular news media. Following the "European year of preservation of monuments," study groups from ETH Zurich investigated in 1975–6 the area of Küçük Ayasofya; Istanbul Technical University started investigating the Süleymaniye and Zeyrek areas. DAI's Istanbul Department, under the leadership of Wolfgang Müller-Wiener, launched in 1976 an ambitious documentation program of the building stock in Zeyrek, continued in 1977 and 1978 in collaboration with Universitat Karlsruhe, TH Darmstadt, and Istanbul Technical University. Many of the houses documented in these research programs were lost in the coming decades, as the findings did not translate to conservation programs. See: Martin Bachmann, "Transitory Worlds: Wooden Houses of Istanbul Portrayed in the Research of the DAI Istanbul," in *Ahşap İstanbul: Konut Mimarisinden Örnekler; [Istanbul-Forschungszentrum Katalog 5] = Istanbuls Holzhäuser: Beispiele Seiner Historischen Wohnarchitektur = Wooden Istanbul: Examples from Housing Architecture,* eds. Martin Bachmann and Baha Tanman (Istanbul: Suna-İnan Kıraç Akdeniz Medeniyetleri Araştırma Enstitüsü, 2008), 109.

27 Güngör Gönültaş, "Avrupa Konseyi Kültür Komitesi 2. Başkanı, Dışişleri Bakanlığı'nın çağrılısı olarak Türkiye'ye geldi. Adno: 'İstanbul'un mimari mirasının korunmasina önemsiyoruz'; eski yapıların korunması icın Soğukçeşme ve Zeyrek pilot bölge seçildi," *Milliyet*, August 3, 1978, 11; Abdullah Öğülmüş, "Avrupa Konseyinin çağrısı ve UNESCO'nun Katkısı sonuç verdi: İstanbul'un 5 semti, 2.5 milyar lira karşılığında 19. yüzyıla dönecek; Dünya Bankası'nın kredisi ile Topkapı, Zeyrek, Süleymaniye, Yenikapı, Yedikule yörelerinde taş kaldırımlar döşenecek, elektrik yerine kandiller ve fenerler kullanılacak, mahallelerde faytonlar dolaşacak," *Milliyet*, July 30, 1979, 7.

28 A UNESCO-sponsored four-year rehabilitation project for Fener-Balat was initiated in 1998.

29 It was again the Ministry of Tourism, which declared Süleymaniye, Eyüp, and Galata as tourism zones in December 1991. Remzi Gökdağ, "Restorasyona Para

Aranıyor: Süleymaniye, Eyüp ve Galata Turizm Bölgelerindeki Tarihi Binalar Elden Geçecek," *Cumhuriyet*, March 9, 1992, 9.

30 Law No. 5366, Yıpranan Tarihi ve Kültürel Taşınmaz Varlıkların Yenilenerek Korunması ve Yaşatılarak Kullanılması Hakkındaki Kanun (Preservation by Renovation and Utilisation by Revitalizing of Deteriorated Immovable Historical and Cultural Properties). "İstanbul Müze Kent Projesi," Governor's Office website, accessed April 28, 2008, http://www.istanbul.gov.tr/?pid=2242. For an early criticism, see: Doğan Kuban, ed., *İstanbul Müzekent Projesi Bağlamında Gözlemler* (Istanbul: Yapı Endüstri Merkezi Yayınları, 2006).

31 On Sulukule, see: Asu Aksoy and Kevin Robins, "Heritage, Memory, Debris: Sulukule, Don't Forget," in *Heritage, Memory and Identity*, eds. Helmut Anheier and Yudhishthir Raj Isar (London: Sage, 2011), 222–30. On Süleymaniye, see: Fatih Pınar and Ayşe Çavdar, "Osmanlılaştırılan Süleymaniye," *Bianet*, April 8, 2009, accessed on January 26, 2017, http://bianet.org/bianet/siyaset/113695-osmanlilastirilan-suleymaniye.

32 Gregor J. Ashworth and John E. Tunbridge, *The Tourist-Historic City* (London: Belhaven Press, 1990), 28; also see, Altınyıldız, "The Architectural Heritage of Istanbul." The first legislation, *Asar-a Atika Nizamnamesi* of 1881, was followed by a revision in 1906. Then came the 1973 law #1710; 1983 law #2863; and amendments in 1987 and 2004.

33 Zeynep Aygen, *International Heritage and Historic Building Conservation: Saving the World's Past* (New York: Routledge, 2013), 26.

34 An important connection between these institutions and Turkey was Turkish conservator Cevat Erder, who headed ICCROM from 1981 to 1988, was a member of the Executive Council of ICOMOS from 1965 to 1974, and founded of the Department of Restoration and Historic Preservation at METU in 1964.

35 The ICOMOS National Committee of Turkey was constituted in 1974. Turkey participated in UNESCO in 1982, in accordance with the "Conservation of World Cultural and Natural Heritage Charter," accessed on April 28, 2008, www.international.icomos.org.

36 Smith, *The Uses of Heritage*, 87.

37 Peter Howard and Gregory Ashworth, *European Heritage, Planning and Management* (Exeter, UK: Intellect Books, 1999).

38 Europa Nostra, accessed on April 28, 2008, www.europanostra.org. For news in local press see, "Önemli Haberler," *Arkitekt*, no. 369 (1978): 3, 16.

39 John Delafons, "European Architectural Heritage Year," in *Politics and Preservation: A Policy History of the Built Heritage, 1882–1996* (London: E & FN Spon, 1975), 110–15.

40 The European Charter of Architectural Heritage, adopted by the Council on September 26, 1975, was announced at the Amsterdam Congress, held on October 21–25, as the Amsterdam Declaration. Rob Pickard and Council of Europe, *Policy and Law in Heritage Conservation I, Conservation of the European Built Heritage Series* (New York: E & FN Spon, 2000).

41 Ibid., 195, 363.

42 Delafons, "European Architectural Heritage Year," 110.

43 Kemali Söylemezoğlu, "Tarihi Çevrenin Oluşumu ve Sinan," *Arkitekt* 44, no. 358 (1976): 51–3; Üstün Alsaç, *Türkiye'de Restorasyon* (Istanbul: İletisim Yayınları, 1992); Zeynep Ahunbay, *Tarihi Çevre Koruma ve Restorasyon* (Istanbul: Yapı Endüstri Merkezi, 2004), 142–3.

44 Bülent Özer, "Mimari Mirası Korumanın Anlamı, Kapsamı ve Sınırları Üzerine," *Yapı*, no. 17 (1976): 33–40; Mehmet Çubuk, "Mimari Miras Yılı İçinde 1975 Amsterdam Kongresi ve Bütünleşmiş Koruma," *Yapı, no.* 14 (1975), 41–3.

45 One such exhibition, a collaboration between Istanbul's Mayor Ahmet İşvan, ITU Architecture Faculty's Architectural History Program, the Touring Club, and the Ministry of Tourism, which originally took place at the Municipality's exhibition build-

ing, was later taken to Strasburg to the European Council. "Önemli Haberler," *Arkitekt*, no. 363 (1976): 99–100.

46 E.g. the exhibition "Profitopolis"—which opened on May 15, 1974, at the Fine Art Academy—and its reviews constitute the first instance (in my survey of *Yapı* and *Arkitekt*) where urban problems were defined as deriving from "profiteering"/speculation and "culture." Originating in the Federal Republic of Germany, and discussing speculative real estate developments there, the exhibition was adapted by a group of local professionals to include a small addition on Istanbul.

47 Haluk Sezgin, "Pullardaki Evler," *Yapı*, no. 9 (1975): 7; Asım Mutlu, "Türk Evleri," *Sanat Dünyamız* 1, no. 3 (1975): 3–13, and "Türk Evi ve Türk Mahalleleri," *Arkitekt*, no. 359 (1975b): 126, 128.

48 Haluk Durukal, "Açıkhava Müzeleri," *Milliyet*, March 16, 1973; and "Bir Serginin Düşündürdükleri: İstanbul'da 'Açıkhava Müzesi' Kurulmalıdır," *Milliyet*, July 12, 1975.

49 Ahmet Köksal, "'İstanbul 1800' Sergisi, Eski İstanbul'u özellikleriyle yansıtacak bir açık hava müzesi öneriyor," *Milliyet Sanat*, no. 135 (1975): 18–20. I highlight this exhibition and not others such as Ruşen Dora's on "Old Houses and Streets" (1975) because it articulated and advocated a proposal for an open-air museum. Ruşen Dora, "Eskimiş Evler ve Sokakları," *Arkitekt*, no. 357 (1975): 19–20.

50 Güngör Gönültaş, "'Eski İstanbul' Sitesi İçin Proje Hazır," *Milliyet*, April 1, 1974, 3; Vasfiye Özkoçak, "1800 yılında İstanbul," *Milliyet Magazin*, July 12, 1975; Selmi Andak, "C.M. Mitchell İstanbul 1800 projesini destekliyor," *Cumhuriyet*, August 20, 1975; Necla Seyhun, "Eski İstanbul: Düşle Gerçek Arasında Bir Kent," *Cumhuriyet*, September 20, 1975, 6.

51 Reşat Ekrem Koçu, "Oya Kılıç'in Maket Sergisi," newspaper clipping. Oya Kılıç Archive.

52 Leyla Erduran, "'İstanbul 1800' İçin," *Milliyet Magazin*, October 24, 1976.

53 For example: Gail Lee Dubrow and Jennifer B. Goodman, eds. *Restoring Women's History through Historic Preservation* (Baltimore, MD; London: Johns Hopkins University, 2003).

54 Aygen, *International Heritage*, 60.

55 My translation. Oya Kılıç, "İstanbul 1800," *Bizler* (1976); Also, see: Oya Kılıç, "Eski İstanbul Açıkhava Müzesi," *Türkiyemiz* (October 1977): 8–14.

56 The work of Metin Sözen merits mention: Sözen worked earlier on the preservation of the town of Safranbolu in Northern Turkey, and in 1990, he founded the Foundation for the Promotion and Protection of the Environment and Cultural Heritage (ÇEKÜL), the leading heritage NGO in Turkey today. However, my account is deliberately focused on the organizations in Istanbul in the 1970s and their inter-connections.

57 Perihan Balcı, *Eski İstanbul Evleri ve Boğaziçi Yalıları* (Istanbul: Doğan Kardeş Matbaacılık, 1975).

58 As told in: *Türk Evi ve Biz* (Istanbul: Türkiye Tarihi Evleri Koruma Derneği Kültür Yayınları, 1993).

59 I had access to facsimile copies of original documents at the TÜRKEV-owned Dede Efendi House. The founding statement was also published in *Haber Gazetesi*, March 18, 1977.

60 Several prominent local scholars, such as Reşat Ekrem Koçu and Doğan Kuban, wrote introductions to Balcı's exhibition brochures. She was one of the contributors to the *Mimaride Türk Milli Üslubu Semineri* (Turkish National Style in Architecture Seminar) of 1984. Doğan Kuban, "Birinci Dünya Savaşı'ndan önce İstanbul Konutları (Bir kesit)," introduction to Perihan Balcı's exhibition "Istanbul'un Eski Ev ve Sokakları (Old Houses and Streets of Istanbul)." Yapı ve Kredi Bankası kültür ve sanat hizmetlerinden 90. sergi, May 3–June 3, 1972; Reşat Ekrem Koçu, "Kaybolan İstanbul (Vanishing Istanbul)," introduction to Perihan Balcı's exhibition "İstanbul'un Eski Boğaziçi Yalıları," January 23–February 5, 1974.

61 Interview with Cengiz Eruzun, "İstanbul nasıl yenileniyor?" *Yeni Mimar*, June 7, 2007.

62 Sinan Genim, "TAÇ Vakfı'nın 25. Yılı," in *TAÇ Vakfı'nın 25. Yılı Anı Kitabı: Türkiye'de Risk Altındaki Doğal ve Kültürel Miras*, ed. Sinan Genim (Istanbul: TAÇ Vakfı, 2001), xv–xvi.
63 Minister of Tourism Lütfi Tokoğlu announced the establishment and funding of TAÇ in March of 1976.
64 This study was published in a special issue on architectural conservation in *Çevre* ('Environment'), a short-lived journal of the built environment produced by ITU and MSU faculty members. Nezih Eldem, Atilla Yücel, and Melih Kamil, "Sultanahmet Meydanı Çevresi ve Ayasofya Soğukçeşme Sokağı Koruma ve Geliştirme Projesi," *Çevre*, no. 3 (1979): 19–23; and Nezih Eldem, Melih Kamil, and Atilla Yücel, "A Plan for Istanbul's Sultanahmet-Ayasofya Area," in *Conservation as Cultural Survival*, ed. Renata Holod (Philadelphia: Aga Khan Award for Architecture, 1980), 53–6.
65 Çelik Gülersoy, "Mesaj." *Sanat Çevresi*, no. 69 (1984). Architects responded in the following publication: *Mesaja Yanıtlar* (Istanbul: TMMOB Mimarlar Odası İstanbul Şubesi, 1984).
66 Caryl Ster, "In Old Istanbul, A New Hotel," *New York Times*, January 19, 1986, 22, accessed January 25, 2017. http://www.nytimes.com/1986/01/19/travel/in-old-istanbul-a-new-hotel.html.
67 The club's income from Customs was sharply cut in 1990 when the central government canceled its contract to charge Turkish workers. After 1994, when the first "Islamist" Municipality of Istanbul, led by the then-mayor Recep Tayyip Erdoğan, asked the club to vacate its leased properties, Soğukçeşme Street remained one of its last bastions of influence.
68 Specifically on Soğukçeşme Street, see: Zeynep Çelik, "Istanbul: Urban Preservation as Theme Park, the Case of Soğukçeşme Street," in *Streets: Critical Perspectives on Public Space,* eds. Zeynep Çelik, Diane Favro, and Richard Ingersoll (Berkeley, CA: University of California Press, 1994), 83–93. On world's fairs and the representation of Muslim peoples via installations such as Cairo Street: Zeynep Çelik, "Islamic Quarters in Western Cities," in *Displaying the Orient: Architecture of Islam at Nineteenth-Century World's Fairs* (Berkeley, CA: University of California Press, 1992), 51–95; Soğukçeşme Street is mentioned on pages 195–6.
69 "İstanbul'da Apartman Olmayan Tek Sokak," *Cumhuriyet*, October 15, 1986, 1.
70 On Ottomanism and collective memory, see: Yılmaz Çolak, "Ottomanism vs. Kemalism: Collective Memory and Cultural Pluralism in 1990s Turkey," *Middle Eastern Studies* 42, no. 4 (2006): 587–602.

Bibliography

Ahunbay, Zeynep. *Tarihi Çevre Koruma ve Restorasyon*. Istanbul: Yapı Endüstri Merkezi, 2004.

Akcan, Esra. *Çeviride Modern Olan*. Istanbul: YKY, 2008.

—. *Architecture in Translation: Germany, Turkey, and the Modern House*. Durham, NC: Duke University Press, 2012.

Aksoy, Asu, and Kevin Robins. "Heritage, Memory, Debris: Sulukule, Don't Forget." In *Heritage, Memory and Identity*, edited by Helmut Anheier and Yudhishthir Raj Isar, 222–30. London: Sage, 2011.

Alsaç, Üstün. *Türkiye'de Restorasyon*. Istanbul: İletisim Yayınları, 1992.

Altınyıldız, Nur. "The Architectural Heritage of Istanbul and the Ideology of Preservation." *Muqarnas* 24, no. 1, History and Ideology: Architectural Heritage of the "Lands of Rum." (2007): 281–305.

Andak, Selmi. "C.M. Mitchell İstanbul 1800 projesini destekliyor." *Cumhuriyet*, August 20, 1975.

Arel, Ayda. *Osmanlı Konut Geleneğinde Tarihsel Sorunlar*. Izmir: Ege Üniversitesi Güzel Sanatlar Fakültesi Yayınları, 1982.

Arseven, Celal Esad. *Constantinople*. Paris: H. Laurens, 1909.

Ashworth, Gregor J., and John E. Tunbridge. *The Tourist-Historic City*. London: Belhaven Press, 1990.

Aygen, Zeynep. *International Heritage and Historic Building Conservation: Saving the World's Past*. New York: Routledge, 2013.

Bachmann, Martin. "Transitory Worlds: Wooden Houses of Istanbul Portrayed in the Research of the DAI Istanbul." In *Ahşap İstanbul: Konut Mimarisinden Örnekler; [Istanbul-Forschungszentrum Katalog 5] = Istanbuls Holzhäuser: Beispiele Seiner Historischen Wohnarchitektur = Wooden Istanbul: Examples from Housing Architecture*, edited by Martin Bachmann and Baha Tanman, 96–202. Istanbul: Suna-İnan Kıraç Akdeniz Medeniyetleri Araştırma Enstitüsü, 2008.

Balcı, Perihan. *Eski İstanbul Evleri ve Boğaziçi Yalıları*. Istanbul: Doğan Kardeş Matbaacılık, 1975.

—. "Eski İstanbul Evleri." *Sanat Dünyamız* 7, no. 21 (1981): 2–12.

Baydar, Gülsüm. "Between Civilization and Culture: Appropriation of Traditional Dwelling Forms in Early Republican Turkey." *Journal of Architectural Education* 47, no. 2 (1993): 66–74.

Bektaş, Cengiz. *Türk Evi*. Istanbul: Yapı Kredi Yayınları, 1996.

Bertram, Carel. *Imagining the Turkish House: Collective Visions of Home*. Austin: University of Texas Press, 2008.

Boym, Svetlana. *The Future of Nostalgia*. New York: Basic, 2001.

Boysan, Burak. "Politik Hummanın Silinmeyen İzleri: Halkla İlişkiler Stratejisi Olarak İstanbul'un İmarı." *İstanbul*, no. 4 (1993): 84–98.

Bozdoğan, Sibel. "The Turkish House Reappraised." In *Sedad Eldem: Architect in Turkey*, edited by Suha Özkan and Sibel Bozdoğan, 44–75. Singapore; New York: Concept Media/Aperture, 1987.

—. "Vernacular Architecture and Identity Politics: The Case of the 'Turkish House'." *Traditional Dwellings and Settlements Review* 7, no. 2 (1996): 7–18.

—. *Modernism and Nation Building: Turkish Architectural Culture in the Early Republic*. Seattle; London: University of Washington Press, 2001.

Brown, Robert, and Daniel Maudlin, "Concepts of Vernacular Architecture." In *The Handbook of Architectural Theory*, edited by Greig Crysler, Stephen Cairns, and Hilde Heynen, 340–55. London: Sage Publications, 2011.

Çelik, Zeynep. *Displaying the Orient: Architecture of Islam at Nineteenth-Century World's Fairs*. Berkeley, CA: University of California Press, 1992.

—. "Istanbul: Urban Preservation as Theme Park, the Case of Soğukçeşme Street." In *Streets: Critical Perspectives on Public Space*, edited by Zeynep Çelik, Diane Favro, and Richard Ingersoll, 83–93. Berkeley, CA: University of California Press, 1994.

Çolak, Yılmaz. "Ottomanism vs. Kemalism: Collective Memory and Cultural Pluralism in 1990s Turkey." *Middle Eastern Studies* 42, no. 4 (2006): 587–602.

Çubuk, Mehmet. "Mimari Miras Yılı İçinde 1975 Amsterdam Kongresi ve Bütünleşmiş Koruma." *Yapı*, no. 14 (1975): 41–3.

Delafons, John. "European Architectural Heritage Year." In *Politics and Preservation: A Policy History of the Built Heritage, 1882–1996*, 110–15. London: E & FN Spon, 1975.

Dora, Ruşen. "Eskimiş Evler ve Sokakları." *Arkitekt*, no. 357 (1975): 19–20.

Dubrow, Gail Lee, and Jennifer B. Goodman, eds. *Restoring Women's History through Historic Preservation*. Baltimore, MD; London: Johns Hopkins University, 2003.

Durukal, Haluk. "Açıkhava Müzeleri." *Milliyet*, March 16, 1973.
—. "Bir Serginin Düşündürdükleri: İstanbul'da 'Açıkhava Müzesi' Kurulmalıdır." *Milliyet*, July 12, 1975.
Eldem, Nezih, Atilla Yücel, and Melih Kamil. "Sultanahmet Meydanı Çevresi ve Ayasofya Soğukçeşme Sokağı Koruma ve Geliştirme Projesi." *Çevre*, no. 3 (1979): 19–23.
Eldem, Nezih, Melih Kamil and Atilla Yücel. "A Plan for Istanbul's Sultanahmet-Ayasofya Area." In *Conservation as Cultural Survival*, edited by Renata Holod, 53–6. Philadelphia: Aga Khan Award for Architecture, 1980.
Eldem, Sedad Hakkı. Türk Evi Plan Tipleri. Istanbul: Istanbul Technical University, 1954.
—. *Türk Evi: Osmanlı Dönemi, Turkish Houses Ottoman Period* I, II, III. Istanbul: Türkiye Anıt, Çevre, Turizm Değerlerini Koruma Vakfı, 1984, 1986, 1987.
Erduran, Leyla. "'İstanbul 1800' İçin." *Milliyet Magazin*, October 24, 1976.
Genim, Sinan, ed. *TAÇ Vakfı'nın 25. Yılı Anı Kitabı: Türkiye'de Risk Altındaki Do al ve Kültürel Miras*. Istanbul: TAÇ Vakfı, 2001.
Gökdağ, Remzi. "Restorasyona Para Aranıyor: Süleymaniye, Eyüp ve Galata Turizm Bölgelerindeki Tarihi Binalar Elden Geçecek." *Cumhuriyet*, March 9, 1992.
Gönültaş, Güngör. "'Eski İstanbul' Sitesi İçin Proje Hazır." *Milliyet*, April 1, 1974.
—. "Avrupa Konseyi Kültür Komitesi 2. Başkanı, Dışişleri Bakanlığı'nın çağrılısı olarak Türkiye'ye geldi. Adno: 'İstanbul'un mimari mirasının korunmasina önemsiyoruz'; eski yapıların korunması icın Soğukçeşme ve Zeyrek pilot bölge seçildi." *Milliyet*, August 3, 1978.
Graham, Brian, Gregory Ashworth, and John Tunbridge. *A Geography of Heritage: Power, Culture and Economy*. London: Oxford University Press, 2000.
Gül, Murat. *The Emergence of Modern Istanbul: Transformation and Modernisation of a City*. London; New York: I.B. Tauris Academic Studies, 2009.
Gülersoy, Çelik. "Mesaj." *Sanat Çevresi*, no. 69 (July 1984).
Günay, Reha. *Tradition of the Turkish House and Safranbolu Houses*, translated by Çelen Birkan. Istanbul: Yapı-Endüstri Merkezi, 1998.
Howard, Peter, and Gregory Ashworth. *European Heritage, Planning and Management*. Exeter, UK: Intellect Books, 1999.
"İstanbul Yılın Şehri." *Hayat*, no. 7, February 12, 1960.
"İstanbul'da Apartman Olmayan Tek Sokak." *Cumhuriyet*, October 15, 1986.
Kılıç, Oya. "İstanbul 1800." *Bizler*, no. 11, Journal of the Building and Credit Bank (1976): 21.
Kılıç, Oya. "Eski İstanbul Açıkhava Müzesi." *Türkiyemiz* (October 1977): 8–14.
Kirshenblatt-Gimblett, Barbara. *Destination Culture: Tourism, Museums, and Heritage*. Berkeley, CA: University of California Press, 1998.
—. "World Heritage and Cultural Economics." In *Museum Frictions: Public Cultures/Global Transformations*, edited by Ivan Karp, Corinne A. Kratz, Lynn Szwaja, and Tomas Ybarra-Frausto, 161–201. Durham, NC: Duke University Press, 2006.
Koçu, Reşat Ekrem. "Ev." In *İstanbul Ansiklopedisi* 10, 5400–406. Istanbul: Koçu Yayınları, 1971.
Köksal, Ahmet. "'Istanbul 1800' Sergisi, Eski İstanbul'u özellikleriyle yansıtacak bir açık hava müzesi öneriyor." *Milliyet Sanat*, no. 135 (1975): 18–20.
Kuban, Doğan. *The Turkish Hayat House*. Istanbul: Muhittin Salih Eren, 1995.
—, ed. *İstanbul Müzekent Projesi Bağlamında Gözlemler*. Istanbul: Yapı Endüstri Merkezi Yayınları, 2006.

Küçükerman, Önder. *Kendi Mekanının Arayışı İçinde Türk Evi*. Istanbul: Türkiye Turing ve Otomobil Kurumu, 1985.
Lowenthal, David. "Identity, Heritage, and History." In *Commemorations: The Politics of National Identity*, edited by John R. Gillis, 41–57. Princeton, NJ: Princeton University Press, 1994.
—. *The Heritage Crusade and the Spoils of History*. Cambridge: Cambridge University Press, 2003.
Mills, Amy. *Streets of Memory: Landscape, Tolerance, and National Identity in Istanbul*. Athens: University of Georgia Press, 2010.
Mimaride Türk Milli Üslubu Semineri. Istanbul: Kültür ve Turizm Bakanlığı Eski Eserler ve Müzeler Genel Müdürlüğü, 1984.
Mutlu, Asım. "Türk Evleri." *Sanat Dünyamız* 1, no. 3 (1975): 3–13.
—. "Türk Evi ve Türk Mahalleleri." *Arkitekt*, no. 359 (1975): 126, 128.
Öğülmüş, Abdullah. "Avrupa Konseyinin çağrısı ve UNESCO'nun Katkısı sonuç verdi: Istanbul'un 5 semti, 2.5 milyar lira karşılığında 19. yüzyıla dönecek; Dünya Bankası'nın kredisi ile Topkapı, Zeyrek, Süleymaniye, Yenikapı, Yedikule yörelerinde taş kaldırımlar döşenecek, elektrik yerine kandiller ve fenerler kullanılacak, mahallelerde faytonlar dolaşacak." *Milliyet*, July 30, 1979.
Özer, Bülent. "Mimari Mirası Korumanın Anlamı, Kapsamı ve Sınırları Üzerine." *Yapı*, no. 17 (1976): 33–40.
Özkoçak, Vasfiye. "1800 yılında İstanbul." *Milliyet Magazin*, July 12, 1975.
Pickard, Rob, and Council of Europe. *Policy and Law in Heritage Conservation* I, Conservation of the European Built Heritage Series. New York: E & FN Spon, 2000.
Pınar, Fatih, and Ayşe Çavdar. "Osmanlılaştırılan Süleymaniye." *Bianet*, April 8, 2009. Accessed on January 26, 2017. http://bianet.org/bianet/siyaset/113695-osmanlilastirilan-suleymaniye.
Pinon, Pierre. "Henri Prost, Albert Gabriel, Istanbul Archeological Park and the Hippodrome." In *Hippodrome / Atmeydanı, A Stage for Istanbul's History*, edited by Brigitte Pitarakis and Ekrem Işın, 152–67. Istanbul: Pera Museum, 2010.
Rajagopalan, Mrinalini. "Preservation and Modernity: Competing Perspectives: Contested Histories and the Question of Authenticity." In *The Handbook of Architectural Theory*, edited by Greig Crysler, Stephen Cairns, and Hilde Heynen, 208–324. London: Sage, 2012.
Salamandra, Christa. *A New Old Damascus: Authenticity and Distinction in Urban Syria*. Bloomington, IN: Indiana University Press, 2004.
Seyhun, Necla. "Eski İstanbul: Düşle Gerçek Arasında Bir Kent." *Cumhuriyet*, September 20, 1975.
Sezer, Yavuz. "The Perception of Traditional Ottoman Domestic Architecture as a Category of Historic Heritage and a Source of Inspiration for Architectural Practice (1909–1931)." MA in History of Art and Architecture. Istanbul: Boğaziçi University, 2005.
Sezgin, Haluk. "Pullardaki Evler." *Yapı*, no. 9 (1975): 7.
Shaw, Wendy M. K. *Possessors and Possessed: Museums, Archeology, and the Visualization of History in the Late Ottoman Empire*. Berkeley, CA: University of California Press, 2003.
Smith, Laurajane. *The Uses of Heritage*. London; New York: Routledge, 2006.
Söylemezoğlu, Kemali. "Tarihi Çevrenin Oluşumu ve Sinan." *Arkitekt* 44, no. 358 (1976): 51–3.
Ster, Caryl. "In Old Istanbul, A New Hotel." *New York Times*, January 19, 1986. Accessed January 25, 2017. http://www.nytimes.com/1986/01/19/travel/in-old-istanbul-a-new-hotel.html.

Tanyeli, Uğur. "Türkiye'de Modernleşme ve Vernaküler Mimari Gelenek: Bir Cumhuriyet Dönemi İkilemi." In *Bilanço 1923–1998: Türkiye Cumhuriyeti'nin 75 Yılına Toplu Bakış Uluslararası Kongresi* 1, edited by Zeynep Rona, 283–90. Istanbul: Tarih Vakfı Yayınları, 1999.

Tanyeli, Uğur, and Bülent Tanju, eds. *Sedad Hakkı Eldem* I, II. Istanbul: Osmanlı Bankası Arşiv ve Araştırma Merkezi, 2008, 2009.

Türeli, Ipek. "Heritagisation of the "Ottoman/Turkish House" in the 1970s: Istanbul-based Actors, Associations and their Networks." *European Journal of Turkish Studies* (online), 19 (2014): 1–27. http://ejts.revues.org/5008.

Türk Evi ve Biz. Istanbul: Türkiye Tarihi Evleri Koruma Derneği Kültür Yayınları, 1993.

Yürekli, Hülya, and Ferhan Yürekli. *Türk Evi: Gözlemler-Yorumlar = The Turkish House: A Concise Re-evaluation*. Istanbul: Yapı-Endüstri Merkezi, 2005.

5 Modeling the city

The Golden Horn (Haliç) estuary suffered over a long period from industrial pollution. Its rehabilitation became an issue as early as the 1950s.[1] In the second half of the 1980s, the government carried out a "cleaning" operation that also involved major and swift demolition of properties crowding its shoreline. Green parks were created on reclaimed land along the shores, but they remained desolate because of the lack of activity. Then the government started encouraging cultural investment around the Golden Horn (see Figure 5.1).

Miniaturk, Turkey's first nation-themed park of miniature models, opened in 2003 and is part of the latter phase of this urban regeneration campaign, which focuses on culture-led revitalization. There had been an earlier proposal in 1989 for an entertainment-oriented (1/33-scale) miniature Istanbul model presented to Istanbul's mayor by renowned architect Cengiz Bektaş, to be located on the nearby Golden Horn (Bahariye) islands, but this proposal was not taken up by the municipality at that first phase of waterfront regeneration.[2]

Since opening, Miniaturk's success has been enormous. It has found a place in the city's popular landscape, and Istanbul's guided tours and printed guides now incorporate it next to well-known historical sites. With keen and enduring support from the press, its total visitor count rose to more than two million by the end of its first two years. Other cities have even tried to build their own versions. For example, the southern tourist city of Antalya built its own "Mini-City," designed by architect Emre Arolat, in 2004.[3] Yet, in contrast to the controversy around other global-city projects, Miniaturk has not triggered any visible opposition.[4] It was designed by architect Murat Uluğ and built on publicly owned land by Istanbul Culture and Arts Products Trade Co. (Kültür A.Ş., Culture Co.), the company that manages the Istanbul Metropolitan Municipality's facilities and activities in the realm of culture. Even though funds for its construction were reportedly acquired from corporate sponsors, as a cultural heritage site it did not appear to benefit any particular private interest.

As its name makes plain, Miniaturk presents a "Turkey in miniature." Its main outdoor display area features 1/25-scaled models of architectural showpieces chosen for their significance in the city's and Turkey's history in distinct adjacent zones. As a site of architectural miniatures, Miniaturk provides an escape from the experience of the everyday. However, it must also be understood in dialectic

Figure 5.1 Relatively recent institutions of culture located on or near the banks of the Golden Horn include: Miniaturk in the Örnektepe neighborhood of the Beyoğlu Municipality (1); Sütlüce (or Haliç) Congress Center, on the site of the city's former slaughter house (2); Rahmi Koç Museum, which focuses on transport, industry, and communications, adapted from Lengerhane, the Ottoman Navy anchor foundry, and the Hasköy dockyard (3); SALT, an exhibition and research venue for art and architecture in the restored building of the former Ottoman Bank building on Bankalar Caddesi (Avenue) in Karaköy (4); Istanbul Museum of Modern Art, in a converted warehouse by the water in Tophane (5); Kadir Has University, adapted from the state-run tobacco depot and cigarette factory (6); Feshane International Fair Congress and Culture Center adopted from the imperial *fez* (hat) factory (7); Bilgi University's campus on the site of the city's first electric power plant, Silahtarağa (8).

relation to "gigantic" new sites of global capital as well as to older sites of nation building.

In this chapter, I examine why a miniature Turkey appeared in 2003 and why it has been received with enthusiasm across the political spectrum. I argue that understanding the place and significance of Miniaturk in the popular historical landscape can shed light on the public reaction to other building projects in Istanbul. The park can also be seen as demonstrating a turning point in Turkish politics, as the "vernacular politics" of Islamism moved to the center, into party politics.[5] Here, Islamism refers to a diversity of outlooks that collectively adhere to the notion that Islam is not private but a public and political matter. Finally, Miniaturk illuminates changing notions of citizenship and national identity in a globalizing city.

The genealogy of an exhibition type

Miniature theme parks abound around the world and reveal much about the contexts in which they are situated. Private enterprises such as Disneyland have become hallmarks of national experience. Other, state-controlled examples may attempt even more explicitly to reconfigure the relationship between a nation and its citizens. Ultimately, the tradition of open-air cultural parks can be traced to world exhibitions that provided a venue for expressions of national identity in the nineteenth century.[6]

According to Timothy Mitchell, world exhibitions were designed to offer Europeans a picture of the world in miniature.[7] Drawing on Martin Heidegger's essay "The Age of the World Picture," Mitchell argues that, in general, the effect of spectacles in the West was "to set the world up as a picture," and so objectify it. The exhibition had a particular place within such a regime of spectacle. It combined "authentic" reproductions of the environments of colonized lands with the real bodies of indigenous residents of those lands in an island-like environment separate from the everyday life of the places in which they were staged. Mitchell also points out that despite attempts to construct the exhibition as a copy, the real world always "presented itself as an extension of the exhibition."[8] There is exhaustive literature on these exhibitions, but their importance as a precedent for miniature environments such as Miniaturk remains relatively unexplored. Not only did the exhibitions contrast the industrial products of the West with the traditional crafts of the East but also the present modernity of the West with its pre-industrial past. Such historical re-creations began at the 1867 Paris Exhibition.[9] A notable later example was the Austrian-sponsored "Old Vienna" at Chicago's 1893 Exhibition; and, not to be outdone, the Germans re-created an "Old Berlin" of the 1650s at the Berlin Industrial Exhibition of 1896. According to Katja Zelljadt, "Old Berlin" in modern Berlin confirmed the modernity of this relatively late-industrializing European city.[10]

The exhibitions formed part of a nineteenth-century exhibitionary complex, which allowed visitors, particularly from the working class, to emulate middle-class manners and internalize an ideal of national citizenship.[11] The typology of the open-air museum, which emerged at the end of the nineteenth century, may be regarded as an effort to turn the temporary environments of the exhibitions into permanent installations. Open-air museums also created permanent spaces where assumed national folkways could be frozen in time. Stockholm's Skansen (1891) is generally cited as the first such open-air museum. It contained real farmhouses, dislocated and reassembled from different parts of Sweden. Supposedly representative of not only different regions but different time periods, the houses were brought to the capital city and placed next to each other to create the image of a timeless, static folk life.[12] To sell the idea of a graspable, nostalgic national identity, Skansen was advertised as "Sweden in miniature." According to Tony Bennett, an important difference between nineteenth-century museums and such open-air museums was that the former excluded the lower classes from its represented narratives, while the latter "work[ed] on the ground of popular memory and restyle[d] it."[13] Another difference is in the relationship between interiors and exteriors. According to Patricia Morton, in world's fairs, the disjunction between exterior and interior had been central to the pedagogy of the exhibition.[14] In contrast, open-air parks present buildings as both interiors and exteriors, especially through tableau, sometimes peopled with mannequins, in the interiors.

Today's miniature parks go one step further: While open-air museums like Skansen seek to simulate a lived environment (generally a disappearing, rural one), the miniature park presents objects that are newly made, with no claim to authenticity. And it presents them merely as exteriors. Ownership and management, scale, materiality, organization, and the layout of objects are other aspects that inform the visitor's experience of a miniature park.[15] In the guise of entertainment, these design variables may be used both to advance a particular political view and inculcate citizens with a sense of national identity.

Contemporary nation-themed miniature parks—such as Miniaturk, Madurodam in The Hague, Beautiful Indonesia Miniature Park (Taman Mini Indonesia Indah) in Jakarta, or Splendid China (Jǐnxiù Zhōnghuá) in Shenzhen—show what their producers (increasingly, state enterprises) think their country ought to be.[16] They seek to represent the nation-state and invite citizens to believe in its benevolence. As demonstrations of ideology, their effect is similar to that of monumental capitol complexes, especially in the modern capital cities of nascent nation-states. Cities such as Ankara and certain monumental building complexes within them aim to display and legitimize the authority of the nation-state. Their "gigantic" scale is intended to socialize citizens, encouraging them to internalize the projected nationalist narrative.[17] By contrast, national miniature parks adopt a different palette of spatial techniques: simulation and miniaturization.

The affects of the miniature and the gigantic are not the same. One early modern thinker who considered this issue was Edmund Burke. Writing in the mid-eighteenth century, he associated the sublime with great objects and with pain-induced admiration; he equated the beautiful with small objects of love and pleasure.[18]

In the current neoliberal era, prestige projects, which aim to attract international investment, approximate Burke's notion of the sublime. A striking example is "Dubai Towers," twin towers with twisting torsos proposed by Dubai Properties for Istanbul's Central Business District in 2005. The publicity campaign for the project elicited animated reaction and public debate, partially on account of full-page advertisements showing twisted versions of everyday objects from computer keyboards to coffee cups. Designed to promote the "innovative" form of the proposed towers, the images of small and painfully twisted objects to stand in for the gigantic form of capital inadvertently and ironically exacerbated public consternation. Eventually, however, the principal investor withdrew from the project as a result of sustained opposition to the municipality's giveaway of state-owned land and lawsuits against the deal.

Prestige projects of neoliberal urbanism where great swaths of publicly owned inner-city land are marketed and sold to international capital without concern for transparency, public participation, or debate may be considered examples of the "gigantic." So may monumental projects of nation building. Projects such as Miniaturk may be considered their antipodes but are nevertheless counterparts.

A national miniature-themed park is not an "easy" miniature to interpret. Susan Stewart's discussion of the miniature and the gigantic is helpful in delineating their relative subjective effects.[19] According to Stewart, the miniature is usually associated with the private collection, while the gigantic is associated with the public. The latter symbolizes authority—i.e. the state, masculinity, and exteriority—and is experienced partially while on the move. The miniature, on the other hand, presents popularity, femininity, and interiority and is experienced as a transcendental space frozen in time. Miniatures, such as miniature railways, are appealing partly because they provoke nostalgia for childhood, preindustrial history, and artisanal rather than alienated labor. However, such a neat dichotomy, as in gigantic versus miniature, is difficult to apply to the case of miniature theme parks.

Like their gigantic counterparts, Miniaturk and similar nation-themed parks are designed to be experienced on the move. Thus, visitors are simultaneously immersed in them, as they walk between models, and privileged with a dominant view from above.[20] Stewart suggests that "the interiority of the enclosed world tends to reify the interiority of the viewer."[21] The miniature park thus may appeal to citizens by helping them imagine their nation in its entirety as an uncontaminated and perfect "island." Within that space of refuge, the miniature theme park empowers its visitor. It invites her to assume the bird's-eye view of the modernist city planner; even though there may be a predetermined path and sequence, the relationship of the model to the visitor suggests that the visitor can edit her own version by spending time around those models that interest her, skipping those that do not, by turning back at times and adopting an alternative sequence. However, since the pathways are social spaces, this "editing" work is always informed by the size and preferences of fellow visitors and the crowds they form.

The design of the park

Miniaturk is located on the bank of the Golden Horn in the Örnektepe neighborhood of the Beyoğlu district municipality. It was initially conceived as part of a local revitalization campaign that concentrates on the Golden Horn.[22] It was also presented by officials as participating in a larger process of establishing Istanbul as a "global brand."[23] With the rise of tourism worldwide since the late 1970s, destinations are increasingly turning to theming to compete with one another.[24] Whole countries are creating images, icons, and advertising campaigns to turn themselves into "destination museums."[25] Marketing requires easily identifiable signs or images, and, toward this end, nation-themed miniature parks may be readily used to market nations. Miniaturk's agenda is thus to provide a marketable image of Turkey, not only to foreign tourists but also to Turkish citizens.

Figure 5.2 After paying the entry fee, visitors are treated to their first overview of the exhibition site from the elevated entry terrace.

Photograph by Ömer Faruk Sen.

Miniaturk's site is a strip of infill land between the water and a motorway (İmrahor Avenue). In terms of design, it aims to abstract itself from its surroundings. In the words of its architect Uluğ, the park deliberately seeks to create a "fairytale-like environment."[26] A tall fence blocks off all external views of the interior, and the entrance complex, with administrative and commercial functions, a restaurant, and a shop, faces a parking lot rather than the street. From here, a carefully controlled entry sequence reinforces the sense of separation. First, a ramp takes visitors to a large, raised terrace over a mini-botanical park. It is only after paying the entrance fee and passing through the entry gates that visitors arrive at a vantage point where Miniaturk is revealed (see Figure 5.2). From this elevated location, visitors can enjoy a view of the entire park from behind a long balustrade, or they can take one of two symmetrical ramps down to the ground-level walkways. Here, the main exhibition area is divided roughly in two by external circular paths, between which are meandering paths where the models are distributed (see Figure 5.3).

Where the ramp to the right (to the side of the motorway) meets the ground/exhibition level, visitors encounter the first model, the Mausoleum of Mevlana (built in 1274 in Konya). According to the park's printed guide, it was

> chosen to be the monument that greets the visitors in Miniaturk because of the love and tolerance we can hear in the call of Mevlana "Come, come again! Infidel, fire-worshipper, pagan / Whoever you are, how many times you have

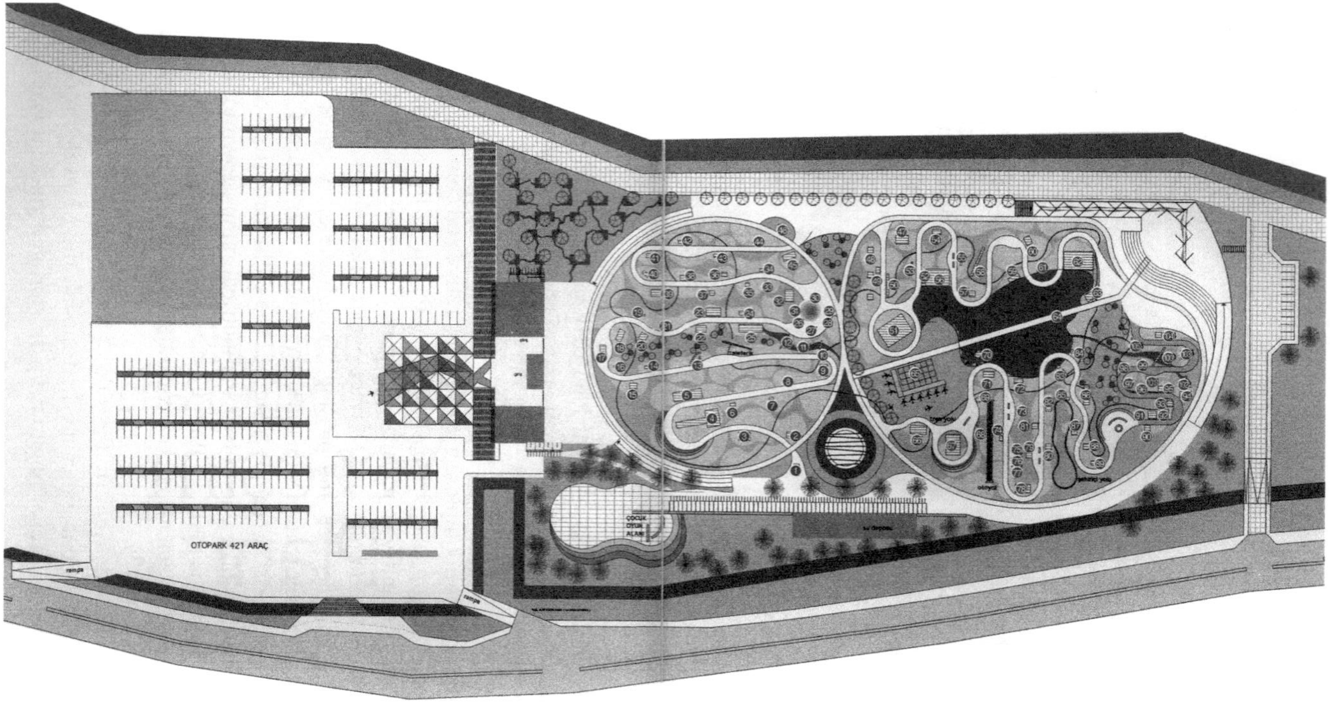

Figure 5.3 Plan published by Miniaturk to solicit models from model-makers.

Miniaturk, Mini Türkiye Parkı Sponsorluk Rehberi, 2002.

sinned, come!" This monument bears witness to the multi-cultural nature of Anatolia.

It is difficult to miss the symbolic message of multiculturalism in the choice of monuments displayed in Miniaturk. The second model is of the Ottoman-era Selimiye Mosque (built between 1568 and 1575 in Edirne); the third is of the Republican-era Anıtkabir (1944–53 in Ankara), Atatürk's mausoleum. The diversity of buildings modeled is also intended to signify the cultural wealth of Turkey. The criteria for selection, visitors are told, are originality and representativeness.[27]

The exhibition space is organized into two main circular areas: the area closer to the entrance terrace contains "Anatolian" sites, while the other contains those from "Istanbul." The Istanbul part is organized around a pond that stands in for the Bosphorus. A model bridge, on which visitors can walk, connects across the pond to an elevated restaurant on the roof terrace of a small building that defines the rear edge of the park. This building is stepped toward the exhibition area to provide outdoor seating and an amphitheater for special events such as concerts. Meanwhile, inside are service areas and a "Panoramic Museum of Victory," which is essentially a diorama in which miniature figures reenact the War of Independence in "Anatolia," complemented by a soundtrack (my analysis here is limited to the outdoor areas).

Outside the two circular exhibition areas, on the Golden Horn side of the park, a third group of models presents a curious selection of buildings dubbed "Ottoman Geography" ("Abroad" is the title used in the English guide). These models—which include the Ecyad Castle in Mecca, the Damascus Train Station, the Dome of the Rock in Jerusalem, and Mostar Bridge in Bosnia-Herzegovina—replicate in miniature buildings outside the current boundaries of Turkey. They were selected, promotional publications state, because they were "built or renovated during the Ottoman Empire." This third group clearly aims to enhance the idea that Turkey not only appreciates multiple cultures within the borders of present-day Turkey, but that it also historically respected and contributed to those cultures once contained within the larger administrative umbrella of the Ottoman Empire.

Overall, the models do not simulate the conditions of original structures in any consistent way: some can be considered miniaturized replicas of what exists on site today, while others are restorations and restitutions of their referents. Most buildings and sites are presented in ways that rid them of their symbolic content; little attempt is made to distinguish between sites based on the communities they serve(d); and some models depict sites that are currently in use, while others are purely historical. Perhaps most startlingly, most of the models are divorced from their original urban contexts. Even buildings located literally next to each other in Istanbul may thus be dislocated and separated in their miniature displays. The most pointed examples of this are from the Dolmabahçe and Sultanahmet areas of Istanbul. In Dolmabahçe, the palace and the clock tower, and in Sultanahmet, the Blue Mosque, Hagia Sophia, Topkapı Palace, and the Cistern, share tightly knit urban contexts. However, in Miniaturk these structures are positioned without reference to each other.

Had the main goal simply been to demonstrate multiculturalism, then Taksim Square would have provided a striking example of the complexity and layered nature of the urban fabric in contemporary Istanbul. On the square are the AKM (Atatürk Cultural Center), Cumhuriyet Anıtı (the Republic Monument, 1928), and the Ottoman water distribution (*taksim*) reservoir (1732), while other significant structures located nearby include the Greek Orthodox Church Ayia Triada (Holy Trinity, 1882) and the French Consulate (formerly the Plague Hospital, 1719). But in Miniaturk, only the Republic Monument (dubbed the Taksim Monument) is reproduced, decontextualized from its context and social significance. By evading all such reference to formal and societal context, the effect is to "naturalize" the originals. Another effect is to create tension between the representation of the Republican and Ottoman eras. However, it is through tensions and contestations that a consensus can be built.

The political context of the park

The literature on theme parks presents these environments as heir to the workings of print in creating the imagination of people as a nation, as described by Benedict Anderson.[28] The nation is mapped for its people in these spaces and reproduced through consumption. However, the designs of nation-themed miniature parks differ in the ways this mapping is materialized. Beautiful Indonesia, for example, orders itself according to official geographical divisions. Thus, each Indonesian province is represented by a pavilion around a central artificial pond, and in the middle of the pond are artificial islands in the shape of the Indonesian Archipelago.[29] Splendid China, on the other hand, arranges its sites to correspond roughly to their real relative locations, but without consistency in scale, while the overall outline of the park resembles China's territorial boundaries.[30]

The officially declared model for Miniaturk is Madurodam in the Netherlands.[31] A comparison, however, reveals that the only real similarity between the two is that both employ building models at 1/25 scale. Madurodam's major claim is that it presents a complete built history of the Netherlands. To accomplish this, it takes the form of a city that has grown outward radially from a medieval core. In Miniaturk there is no formal reference to the map of the nation-state, nor was the possibility of growth designed into the site plan. Only seventy-five models were listed in the park's 2003 visitor's guide (thirty-six from Istanbul, thirty-one from Anatolia, and eight from "abroad"), but 105 were listed in its 2010 website.[32] Additions over time have thus increased the total number and density of the models and transformed the spatial experience of the exhibition space, making it denser. However, its overall boundaries and original thematic groupings remain.

The decision to build a bounded miniature park in which the nation could be viewed in its entirety may be seen in part as a response to anxieties about Turkey's future. As elsewhere, national unity has long been a source of collective paranoia in Turkey. Following World War II, fears for national sovereignty were fueled by

Cold War politics, and since the early 1990s, separatist movements and violent developments in the Balkans and the Middle East have again exacerbated such fears. As a result, the rise of Islamist and Kurdish movements at home has been regarded as threats to national sovereignty, rather than a call for social justice and democracy.

Two principal observations reinforce the notion that Miniaturk may be responding to such anxieties. At the level of image, Miniaturk stage-manages history by painting an ambiguous picture of societal harmony. At the level of production, it attempts to provide a showcase for both the quality and effectiveness of the local and central governments' mutual political vision for the country and the city.

The version of history presented at the park is open to different readings according to the visitor's "viewing position," influenced as this may be by such factors as personal history and the political orientation. Those involved in creating the park may also have interpreted its purpose dissimilarly. Ultimately, its form has embodied negotiation between a range of politicians, administrators, designers, engineers, builders, consultants, and sponsors. Yet, despite these interpretive ambiguities, Miniaturk has been successfully promoted through advertising and word of mouth. Stories have been written in local magazines and newspapers; articles have appeared in tourist and architectural guides to the city; and images of the park have been posted on billboards (also managed by Kültür A.Ş.).

With so much coverage and so many actors, it is understandable that certain discrepancies surfaced in terms of credit for its design and realization. This eventually created a need to stage multiple "opening ceremonies." Miniaturk was actually opened three times: first, on April 23, 2003; second, by (former) Prime Minister (current President) Erdoğan on May 2, 2003; and finally, by Mayor Müfit Gürtuna on May 29, 2003, for the commemoration of the 550th anniversary of Istanbul's "conquest."[33] Meanwhile, the architect is not usually named in official publications of Miniaturk. However, he has published his design drawings in local professional journals, without much reference to the political actors who commissioned the project.

Significantly, the first opening coincided in the official calendar of the Republic with "National Sovereignty and Children's Day." By conflating national sovereignty with a celebration of the child, the date not only implied that the preservation of sovereignty should be a concern of children, but it cast adult citizens in the role of children in relationship to the state. By comparison, Istanbul's conquest commemoration date, May 29 (the third opening date), is not listed in the official calendar of the Republic but nonetheless is celebrated at the local level with minimum official participation. Instead, the marketing of Miniaturk as a conquest-commemoration project by then-Mayor Gürtuna must be viewed in relation to public discussions and competing claims on national history.

As Republican history is commemorated via days such as April 23, May 19 (Atatürk Commemoration, Youth and Sports Day), and October 29 (Republic Day), Islamist civil society groups and Islamism-inspired political parties felt a need to establish alternative days commemorating important events from Turkey's Ottoman past. The counterpart to these commemorations has been a growing nostalgia for the early years of the Turkish Republic among citizens who identify with "Kemalizm" (after Mustafa Kemal Atatürk) and his doctrine of secularism. For example, Esra Özyürek and Kimberly Hart observe the "miniaturization," commodification, and consumption of Atatürk insignia among a variety of strategies through which the early Republic is remembered and politics is privatized in line with neoliberal symbolism of privatization and market choice.[34] According to Özyürek, Kemalist and Islamist versions of the early Republic in the 1990s competed but provided comparably homogenous, dominant narratives.[35]

The backdrop for competing programs of commemoration involved the reluctance of official nationalist history to recognize the accomplishments of the Ottomans. Especially in its early decades, the Republic attempted to establish itself as a secular, West-oriented nation-state, in historic opposition to the old, Islamic, "Eastern" Ottoman Empire. By downplaying conquest-commemoration festivities, it has also sought to avoid upsetting the "European club" by conjuring memories of a past when the Ottoman Turks were a rival power in Europe. For their part, Islamists in government have turned to neo-Ottomanist nostalgia precisely because of, and as a reaction to, the secular Turkish establishment. The Ottomans had self-consciously promoted a multicultural society as well as claiming the leadership of Islam. In contrast, the Republican era (1923–50) was characterized by disrepudiation of the public display of religion. The nascent Turkish state promoted civic nationalism, manufacturing homogeneous national identity out of a very diverse population, but ended up imposing the language (Turkish) and the religion (Sunni Islam) of the majority. [36] Accordingly, citizenship emphasized the individual's duties to the nation-state over his or her rights. [37] However, market reforms in the 1980s, political liberalization in the 1990s, and the EU membership process at the beginning of the 2000s have all contributed to a reevaluation to accepted norms of citizenship, national identity, and the assumed correspondence between them.

This relationship between the state and the citizen is perhaps nowhere more spatially expressed than in the siting and design of the Anıtkabir. On Atatürk's death, his mausoleum was constructed on an imposing hill crowning the capital city of Ankara. Various features of the building sought to embody and define early Republican ideals about the correct relationship between citizen and nation-state.[38] In the years since it was built, the Mausoleum has served as Turkey's official nationalist pilgrimage site. It is where state ceremonies are held, and private and public associations gather there to pay their respects to Atatürk and display their commitment to protect the nation. In the 1990s protests also took place there against such perceived threats to the secular establishment as the headscarf.

One result of the rise of Islamism in the public sphere in the 1990s has been to challenge the state-sanctioned narrative of national identity, and the need to give physical representation to these forces has resulted in efforts to rediscover Istanbul's architectural and urban history as the former Ottoman capital. Alev Çınar argues that the substitute narratives of nationhood produced by Islamist political organizations sought to cast Istanbul as a "victim" of the Republic—thus the reenactment of its conquest symbolically served to "save" it.[39] She explains that in 1994 the Istanbul Metropolitan Municipality under Refah (Welfare) Party and an Islamist nongovernmental organization, the National Youth Foundation, jointly organized the conquest celebration as an event that would rival national commemoration days in scope and scale (more on this in Chapter 6; and also photos of these reenactments).

Among other things, this drew public and academic attention to Islamist claims to public spaces in the city. The victory of Welfare Party in the local elections in major cities, and specifically Erdoğan's in Istanbul, raised further alarms. When Welfare Party leader Necmettin Erbakan became Turkey's first Islamist Prime Minister in 1996, concerns mounted even further. Some even asked if Turkey was set to turn into a new Iran. In 1997, the army, which sees itself as the guardian of the secular Republic, intervened and ousted Welfare Party.[40] The constitutional court then closed down both Welfare Party and its successor, Fazilet (Virtue) Party. These moves by a coalition of the army and other secular elites to block the rise of Islamists swayed electoral support to a reformist faction within. In 2002 this faction, now represented by the AKP, won the general elections under the leadership of Istanbul's prominent former ex-mayor and Turkey's current President Erdoğan, in part by promoting an economic program of "communitarian-liberal syntheses."[41] Back then, in response to questions regarding his new "reformed" position, Erdoğan defined himself not as an Islamist but as a "conservative democrat." Although he used Islamist devices in his daily performance as the prime minister, he made it clear that he considered religion a private issue and that the AKP would not conflate religion and government, Islam and democracy, and would operate within the nation-state's constitutional framework.[42] The AKP subsequently used the European Union membership accession process to negotiate potential threats to it from the secularist establishment. Initially, it advocated a much milder adherence to religion than its predecessor and claimed it recognized a need for societal plurality.[43]

A ground of conversation?

In representing cultural "wealth," Miniaturk seeks to build consensus from contestation. In its version of Turkey, it does not single out the Islamic-Ottoman past but overtly privileges Istanbul over Ankara, the capital built to showcase nation building in the early Republican era. As a reflection of this bias, the models at Miniaturk aim to represent major religious communities that cohabited the land—in line with what is today perceived as Ottoman "cosmopolitanism."

One of the forms of neo-Ottomanist nostalgia that manifests itself most strongly, then, is a longing for the multi-religious and multilingual composition of the city, with its Christians, Jews, and Europeans.[44] For Islamist organizations and individuals, this expresses Muslim hegemony over others. For others, it demonstrates the cultural richness and tolerance of Turkey and its suitability to the new Europe. Miniaturk projects the message of multiculturalism by its selection of Ottoman-era monuments of Muslims and non-Muslims, and pre-Ottoman sites from within the boundaries of Turkey. The incorporation of pre-Ottoman sites from Anatolia, especially those of antiquity, into national patrimony, is very much in line with Republican-era history writing. Yet, the selection constrains those from the Republican era by limiting their number, while the design of individual monuments and their siting negate all societal significance; this is especially true of Republican-era monuments, some of which have immense political significance. The inclusion of monuments from "Ottoman geography" outside of Turkey suggests the park extends the imagination of the nation-state beyond the current borders to claim the Ottoman Empire as historical precedent and cultural heritage. This reflects the current government's (nostalgic) desire to extend Turkey's sphere of influence to that territory once more. The contradictions in design credits and inaugural dates notwithstanding, the embrace of the park by the municipality, the visiting public, and the news media all suggest that it does facilitate a ground for conversation.

In a meeting with one of the model-makers and a public relations representative, I was intrigued to discover how the two identified with different versions of national history, even while working together. One of my questions was directed at the model-maker.[45] When I asked specifically which his favorite model was, he replied that it was the Mağlova Aqueduct (1554–62), and gave the following reasons:

> *Model-Maker*: Not because it is a good model; in fact, it was one of our first. … I like it because of the work itself. It shows they had the determination, the belief, and the will to work. I am not saying this because of indoctrination: "Ottomans, Oh! Ottomans. …" However, every society, like an organism, has a period in which it is alive. … For example, if we bring the first and last decades of the Republic together … [Laughs].
>
> *Public Relations Representative*: Not even bringing the last century together would suffice. … Once, I was leading a group of journalists [in Miniaturk]. One asked why there aren't many examples from the Republic. And I pointed to the squatter settlements [overlooking the park] and said, "There!" He started exclaiming, "You are indeed Ottomanists, you are this … and you are that. …" His own cameraman reacted. "Hold on a minute," he interrupted: "[as if] we have them [Republican landmarks worthy of display], and it is Miniaturk that does not display them?"
>
> *Model-Maker*: But they [squatter settlements] are not the result of the Republic. They exist because of globalization, because of the conjuncture, because of the Cold War.
>
> *Public Relations Representative*: They are the result of a homogenizing world. Look at a plaza. Is this building in New York, Paris, in Istanbul, or in Tokyo? One cannot tell. All look the same. But look at a structure from the Middle Ages, and you can tell at a glance where it's from. If one is equipped with some historical knowledge, one can even identify the country where it is from. But plazas do not allow this.[46]

In this partial exchange, the model-maker conflates the copy (model aqueduct) with the original (real aqueduct). He does not refer to the original as a functional architectural object or space, but as a symbol of technological superiority, as an artifact of cultural self-confidence. The model-maker also suggests that society is an organism, and that architecture is its reflection. Thus, when an organism reaches maturity, it produces monumental architecture and engineering. The Republic, in such a narrative, becomes a period of delayed replenishment that has been terminated by more powerful processes such as internationalization and globalization that have come from without.

For the public relations representative, however, the organism analogy does not work. Her understanding evolves more along an axis of tradition versus modernity. Once there was a time of heterogeneity, but this was defeated by modernization (which started with the Ottoman reforms in the nineteenth century, but was pursued forcefully under the Republic through the twentieth). She thus externalizes the Republic as an agent of top-down modernization. In this analysis, processes of modernization ultimately dictate cultural homogeneity, and Miniaturk becomes a project of "resistance" because it reinstates heterogeneity.

Clearly, the model-maker and the public relations representative exhibit different ideas about Ottoman/Turkish history and how the present built environment has come into being. But they parallel each other in their understanding that architecture is an outcome (rather than, for instance, a cause or an agent) of social processes. Thus, while they do not agree at an ideological level, they have been able to work together to produce representations of architecture that not only display but prove Turkey's exceptionality. In essence, then, Miniaturk offers a common ground for two people who do not share the same conception of history, but who have similar anxieties about cultural homogenization and decline. The exchange between them illustrates the park's potential to reveal political differences, just as it facilitates their concealment.

Erdoğan and his senior aides in the Istanbul Metropolitan Municipality, including the mayor of Istanbul, are eager to shoulder the global-city project and have been proud to initiate and build prestige projects—referred to as *plaza*s in the previous exchange—and to "market" Istanbul to international investment. This administration is essentially continuing a project which begun under Motherland Party with Turgut Özal as Prime Minister and Bedrettin Dalan as Istanbul's mayor (1984–89). The restructuring of the city's economy initiated with the touristification of the city's heritage with a renewed official focus on its Ottoman past and the opening of numerous five-star hotels on prime sites overlooking the Bosphorus. Through the 1990s and 2000s, the city acquired numerous new hotels, gated communities, high-rise condominiums, shopping malls, entertainment venues, and a new skyline of high-rise buildings, especially in the new Central Business District, which now characterizes the city's contemporary face.

Miniaturk has also been used to market the city, but its design and status as a newly built heritage site seem to set it in a different orbit. Thus, when the public-relations representative refers to the homogenization of the built environment via *plaza*s, she assumes that Miniaturk somehow resists this process. What she chooses not to see is that all projects involving prime sites offered to global capital or leisure environments such as Miniaturk are characteristic architectures of globalization. Both are domestic translations of global types; both entail the rerouting of public sources into private or privatized services; and both are designed to demonstrate Turkey's competitiveness in the global marketplace.

Miniaturk reflects a desire to imagine a Turkey that displays its cultural wealth and influence with confidence and pride while being clearly bounded and secure. One image that Miniaturk repeatedly uses in promotional publications (possibly to represent its inclusive politics) shows Atatürk's Mausoleum and the Selimiye Mosque together. The coupling brings into mind the controversy around the Mausoleum and the new Kocatepe Mosque in Ankara (modeled on several Imperial Ottoman mosques, including Selimiye), as discussed by anthropologist Michael E. Meeker. In reality, the Mausoleum and the Kocatepe Mosque stand at similar elevations and crown the two highest hills of the capital city. They thus represent competing claims to Turkey's national imaginary—the Mausoleum standing for secular modernism, the mosque for a modern Islam. This "controversy" is complicated today by the fact that army generals who visit the Mausoleum to pay tribute to Atatürk also perform the *namaz* in Kocatepe; and when the AKP leaders visit the Mausoleum, their members with headscarves will stay out of sight. Both sides display reluctance but simultaneous consensus.

The composition of the Miniaturk promotional photograph is complex in symbolic terms and yields two very different readings. According to one, the Mausoleum, the symbol of Republican nationalism, is clearly foregrounded. But the position of the Mausoleum model on flat ground can alternatively be seen as negating its elevated position in relation to the capital city and the nation. The photograph also shows a few spectators giving a passing glance at the Mausoleum while many more crowd around the Selimiye Mosque (see Figure 5.6). Could the implication be to point out that the Ottoman past is becoming increasingly attractive, in contrast to the Republican one? The angle of view and foreshortening in the image further contributes to this second reading by locating the mosque above the Mausoleum on the printed page. I am not suggesting that the photographer took this image with a definitive discursive claim; but the choice as publicity image suggests the photograph presents the institutional goal of Miniaturk, of building consensus from contestation.

The embodied experience of the theme park by the park's visitors does not privilege any static point of view. The layout of routes through the park is based on an assumption that visitors will stay for two hours.[47] During this time, security personnel and visual and audio messages repeatedly remind them to avoid walking on the grass or touching the models. Meanwhile, a specially commissioned musical composition by Fahir Atakoğlu is broadcast from disguised speakers, and detailed information on individual landmarks is provided in a booklet, as well as through an audio information system activated by the individual user (and again heard through disguised speakers).

The lack of shade, lack of seating, narrow width of paths, and admonishments against touching the models (or even getting close to them) all lead visitors to

Figure 5.4 The only model visitors can touch and walk on is that of the Bosphorus Bridge, which crosses the artificial pond on the site.

view the models while in motion. Only on the model of the Bosphorus Bridge are visitors treated somewhat differently (see Figure 5.4). The actual suspension bridge connects Asia and Europe and is traversed only by motor vehicles. But here it serves as a finale of sorts, raising visitors from the ground level to the elevated restaurant terrace, from where they can look back over the park and contemplate the miniature world that has just been presented to them.

Seeing the city and nation in miniature

The models in Miniaturk are devoid of social life, isolated from any urban context, and laid out in a seemingly arbitrary manner. Instead of trying to depict an overt normative Turkish or Islamic-Turkish character, the park's main rhetorical purpose thus seems to be to indicate the tolerance of Turkey toward multiple cultures and ways of life. Miniaturk's final size has been predetermined; it will not allow for extension or change. It offers the possibility of grasping the entire nation as if it were an island separated from everyday reality and history. It provides a space of refuge in which to imagine a nation without conflicts—one whose sphere of influence reaches, nonetheless, to lands once under Ottoman rule.

Visitors understand and judge the park according to their personal reference systems. They may also understand Miniaturk's inclusivity as a form of nostalgia. But, in the end, their appreciation of the park, demonstrated by attendance numbers and positive reviews, reflects a yearning for alternative modes through which to imagine the nation. As the ethos of Republican nationalism fades away, other sites have emerged to challenge Atatürk's Mausoleum as the center of national symbolism. As the progressivism and secularism of the early Republican nation building project is increasingly criticized from within and a new plurality emerges in its place, Turkey's archive of official symbolic objects and narratives is due for renovation through additions that cater to a new polity. There are many, potentially conflicting aspects of this new polity: the recognition of religion; a renewed interest in Istanbul as the potential gateway through which Turkey will join the multicultural European Union; a nostalgia for Ottoman cosmopolitanism; a drive for the bourgeois beautification of the city; and, finally, a reconfiguring of the relationship between the state and its citizens in the midst of growing dissent towards the representational quality of the democratic process. Because Miniaturk seeks to fulfill all these criteria in a Turkey striving to reassert itself as one among equals in a globalizing world, it has potentially become a new nationalist pilgrimage site. It is in such contexts that discrepancies between people can be willingly suppressed, and memory, accordingly, stylized.

Miniaturk privileges Istanbul, giving it almost equal space in its representation as it does to the rest of the national territory. This condition reflects an imagination that seeks to attach to a global world via Istanbul. While Istanbul is positioned as a center of cultural imagination, it is detached from the imagery of the nation. Likewise, the nation is left with a displaced center. In trying to represent an ideal of Istanbul, the park also tries to detach itself from the reality of the city around it.

Figure 5.5 Miniaturk park: even though the design of the park aims to abstract itself from its surroundings, the surrounding topography impinges in the form of nearby hillside apartment buildings that create an unavoidable backdrop to the models—the source of the public relations representative's disturbance.

Miniaturk: Turkey's Showcase, 2003.

Figure 5.6 Miniaturk's 2003 brochure shows, from left to right, the diagrammatic location of the park in the city, an unstaged snapshot of visitors in the park, and a staged shot of an ideal family behind the Byzantine church-turned-mosque-turned-museum of Hagia Sophia under the words, "The Showcase of Turkey."

Miniaturk's 2003 brochure.

Figure 5.7 "Historic Opening in Europe!" The title of this double-page advertisement in a Turkish newspaper exclaims to advertise a new housing development, "Bosphorus City," and its theme of the Bosphorus surrounded by "*yalıs*."

The models presented in the Istanbul section showcase not only old monumental buildings but also new infrastructural ones. While touching the models is prohibited, visitors are encouraged to walk across the pond that stands in for the Bosphorus on the model of the Bosphorus Bridge. They symbolically enact the (hackneyed) metaphor of crossing the bridge to Europe. Furthermore, as they walk on predetermined pathways around the decontextualized 1/25-scale models, the visitors still have the opportunity to determine how long to dwell on each model, to go back and forth as desired; their bird's-eye view allows to them to temporarily assume the modernist planner's view and create their alternative narrative about the city. Yet, because of the topography of the surrounding terrain, the city becomes an unwanted backdrop to the whole exhibition and undermines its island-like quality (see Figure 5.5). This contrast between the real city and the imagined sterile one, however, may be fundamental to the rhetorical power of the theme park, especially as themed housing developments proliferate on the edges of the city (see Figure 5.7). Indeed, in the past decade, the popularity of a miniaturized Istanbul has moved on from the representational realm of the exhibition to the world outside of it. *à la* Baudrillard, Miniaturk "exists in order to make us believe that the rest is real" while some of what is being built outside "belongs to… the order of simulation."[48]

Notes

1 "Haliç'in temizlenmesi icin proje hazırlanıyor," *Yeni İstanbul*, January 30, 1957.
2 "Haliç adalarına İstanbul maketi," *Cumhuriyet*, December 20, 1989, 7.
3 "Minicity açıldı," *Arkitera.com*, May 31, 2004, accessed May 27, 2017, http://v3.arkitera.com/v1/haberler/2004/05/31/minicity2.htm; "Batman'a 'Miniaturk' yapılacak," *Arkitera.com*, July 21, 2003, accessed May 27, 2017, http://v3.arkitera.com/v1/haberler/2003/07/21/batman.htm.
4 Prestige projects—such as the Haydarpaşa Port, Galataport, and the Dubai Towers, in which prime, publicly owned sites in Istanbul were designated for privatization—were all met with criticism in the media and were resisted by civil society organizations. All three of the projects mentioned here were announced in 2005, and there has been extensive coverage of them in Turkish media. By contrast, there were only a few voices raised against public spending on a "new" site such as Miniaturk, while "real" heritage nearby awaits reinvestment in decrepitude. See Oktay Ekinci, "Dünya Mirasında 'Miniaturk'! . . . ," *Cumhuriyet*, June 15, 2003. Republished in Oktay Ekinci, *İstanbul'un "İslambol" On Yılı* (Istanbul: Anahtar Yayınları, 2004), 53–5.
5 Jenny B. White, *Islamist Mobilization in Turkey: A Study in Vernacular Politics* (Seattle, WA: University of Washington Press, 2002).
6 Tony Bennett, *The Birth of the Museum: History, Theory, Politics* (London: Routledge, 1995); Zeynep Çelik, *Displaying the Orient: Architecture of Islam at Nineteenth-Century World's Fairs* (Berkeley, CA: University of California Press, 1992); and Edward N. Kaufman, "The Architectural Museum: From World's Fair to Restoration Village," *Assemblage* 9 (1989): 20–39.
7 Timothy Mitchell, *Colonising Egypt* (Berkeley, CA: University of California Press, 1991), 1–33.
8 Ibid.
9 Çelik, *Displaying the Orient*, 142–3.
10 Katja Zelljadt, "Presenting and Consuming the Past, Old Berlin at the Industrial Exhibition of 1896," *Journal of Urban History* 31, no. 3 (March 2005): 306–3.

11 Bennett, "The Exhibitionary Complex," in *The Birth of the Museum*, 59–88.
12 Official website of the "Open Air Museum," http://www.skansen.se/eng/, accessed Nov. 29, 2005. Also see Mark Sandberg, *Living Pictures, Missing Persons: Mannequins, Museums, and Modernity* (Princeton, NJ: Princeton University Press, 2003).
13 Bennett, *The Birth of the Museum*, 118.
14 Patricia Morton, *Hybrids of Modernities: Architecture and Representation at the 1931 Colonial Exposition, Paris* (Cambridge, MA: The MIT Press, 2000).
15 Shelly Errington, *The Death of Authentic Primitive Art and Other Tales of Progress* (Berkeley, CA: University of California Press, 1998), 199.
16 Ann Anagnost, *National Past-Times: Narrative, Representation, and Power in Modern China* (Durham, NC: Duke University Press, 1997).
17 Lawrence J. Vale, *Architecture, Power, and National Identity* (New Haven, CT: Yale University Press, 1992).
18 Edmund Burke, *A Philosophical Enquiry into the Sublime and Beautiful* (1757), ed. James T. Boulton (London: Routledge Classics, 2008).
19 Susan Stewart, *On Longing: Narratives of the Miniature, the Gigantic, the Souvenir, the Collection* (Durham, NC: Duke University Press, 1993).
20 Errington, 194–8.
21 Stewart, 64.
22 Dikmen Bezmez, "The Politics of Urban Waterfront Regeneration: The Case of Haliç (the Golden Horn), Istanbul," *International Journal of Urban and Regional Research* 32, no. 4 (2008): 815–40.
23 Cengiz Özdemir, *Faaliyet Raporu 2002* (Istanbul: Kültür A.Ş., 2003), 8.
24 Mark Gottdiener, *The Theming of America: Dreams, Media Fantasies, and Themed Environments* (Boulder, CO: Westview Press, 2001).
25 Barbara Kirshenblatt-Gimblett, *Destination Culture: Tourism, Museums, and Heritage* (Berkeley, CA: University of California Press, 1998).
26 Murat Uluğ, "Miniaturk," *Yapı*, no. 262 (2003): 71–75. Also, Uluğ, "Miniaturk," *Arredamento Mimarlık*, no. 139 (September 2001): 38–39. Author's interview, Nov 2, 2005.
27 The official explanation regarding the selection of landmarks represented in the park states: "Works that found a place in Miniaturk were ones that displayed peculiarities of the era in which they were built, ones that reflected the culture and art of a land that had witnessed thousands of years of heavy invasion, war and destruction, works that had not been destroyed simply because they had been created by those who came before, works that were protected, repaired and enjoyed." "The Selection of Models," *The Showcase of Turkey, Miniaturk, The Story of How It Came to Be* (Istanbul: Kültür A.Ş., 2003), 27.
28 Benedict Anderson, *Imagined Communities: Reflections on the Origin and Spread of Nationalism* (London: Verso, 1991); and Anagnost, *National Past-Times*.
29 For a discussion of Taman Mini, see Michael Hitchcock, "Tourism, Taman Mini, and National Identity," *Indonesia and the Malay World* 26, no. 75 (1998): 124–35; John Pemberton, "Recollections from 'Beautiful Indonesia': Somewhere Beyond the Post-Modern," *Public Culture* 6, no. 2 (1994): 241–62; James T. Siegel, *Fetish, Recognition, Revolution* (Princeton, NJ: Princeton University Press, 1997), 3–5; Abidin Kusno, *Behind the Postcolonial: Architecture, Urban Space, and Political Cultures in Indonesia* (New York: Routledge, 2000), 74–78; and Errington, *The Death of Authentic Primitive Art*, 194–8.
30 For a discussion of Splendid China, see Anagnost, *National Past-Times*.
31 Administrators of Madurodam served as consultants to the creators of Miniaturk. See Musa Ceylan interview with Peter Verdaaddank, "Miniaturk Excites Us," *Gezinti*, Summer 2003, 28–31.
32 www.miniaturk.com.tr/.
33 Özdemir, *Faaliyet Raporu 2002*, 13.
34 Kimberly Hart, "Images and Aftermaths: The Use and Contextualization of Atatürk Imagery in Political Debates in Turkey," *Political and Legal Anthropology* 22, no. 1

(1999): 66–84; and Esra Özyürek, *Nostalgia for the Modern: State Secularism and Everyday Politics in Turkey* (Durham, NC: Duke University Press, 2005).

35 Esra Özyürek, "Public Memory as Political Battleground: Islamist Subversions of Republican Nostalgia," in *The Politics of Public Memory in Turkey*, ed. Esra Özyürek (Syracuse, NY: Syracuse University Press, 2007), 114–37.

36 Thomas W. Smith, "Civic Nationalism and Ethnocultural Justice in Turkey," *Human Rights Quarterly* 27, no. 2 (2005): 436–70. There is extensive literature on nationalism in Turkey.

37 Ayşe Kadıoğlu, "Citizenship and Individuation in Turkey: The Triumph of Will over Reason," Cemoti, no. 26 (1998), accessed March 28, 2006, http://cemoti.revues.org/; E. Fuat Keyman, and Ahmet İçduygu, "Citizenship, Identity, and the Question of Democracy in Turkey," in *Citizenship in a Global World: European Questions and Turkish Experiences*, eds. Fuat Keyman and Ahmet İçduygu (London: Routledge, 2005), 6.

38 Michael Meeker, "Once There Was, Once There Wasn't: National Monuments and Interpersonal Exchange," in *Rethinking Modernity and National Identity in Turkey*, eds. Sibel Bozdoğan and Reşat Kasaba (Seattle, WA: University of Washington Press, 1997), 157–91.

39 Alev Çınar, "National History as a Contested Site: The Conquest of Istanbul and Islamist Negotiations of the Nation," *Comparative Studies in Society and History* 43, no. 2 (2001): 365. Also see Çınar, *Modernity, Islam, and Secularism in Turkey: Bodies, Places, and Time* (Minneapolis, MN: University of Minnesota Press, 2005).

40 Ümit Cizre and Menderes Çınar, "Turkey 2002: Kemalism, Islamism, and Politics in the Light of the February 28 Process," *The South Atlantic Quarterly* 102, no. 2/3 (2003): 309–32.

41 E. Fuat Keyman, and Ahmet İçduygu, 16.

42 Smith, "Civic Nationalism and Ethnocultural Justice in Turkey."

43 Elected for a third term in 2011, however, the AKP became a one-party democracy, crushing opposition. Only by 2012 did Western media started reporting on the curtailing of the freedom of speech in Turkey. One reason has been the United States' need for a stable "Islamic" ally in the region. Dexter Filkins, "Turkey's Jailed Journalists," *The New Yorker*, March 2, 2012, accessed May 27, 2017, http://www.newyorker.com/news/daily-comment/turkeys-jailed-journalists; and "The Deep State," *The New Yorker*, March 12, 2012, accessed May 27, 2017, www.newyorker.com/magazine/2012/03/12/the-deep-state; Fiachra Gibbons, "Turkey's Enlightenment Languishes, Like the Journalists in its Prisons," *The Guardian*, March 13, 2012, accessed May 27, 2017, https://www.theguardian.com/commentisfree/libertycentral/2012/mar/13/turkey-enlightenment-journalists-prisons.

44 The academic literature problematizing and discussing the limits of various versions of Ottomanism is vast. The following references are exemplary: Christopher Houston, *Islam, Kurds and the Turkish Nation State* (Oxford: Berg, 2001); Christopher Houston, "The Brewing of Islamist Modernity: Tea Gardens and Public Space in Istanbul," *Theory, Culture and Society* 18, no. 6 (2001): 77–97; Benton Jay Komins, "Depopulated Cosmopolitanism: The Cultures of Integration, Concealment, and Evacuation in Istanbul," *Comparative Literature Studies* 39, no. 4 (2002): 360–85; Amy Mills, "Gender and Mahalle (Neighborhood) Space in Istanbul," *Gender, Place & Culture* 14, no. 3 (2007): 335–54; and Marcy Brink-Danan, "Counting as European: Jews and the Politics of Presence in Istanbul," in *Orienting Istanbul: Cultural Capital of Europe?*, eds. Deniz Göktürk, Levent Soysal, and Ipek Türeli (New York; London: Routledge, 2010), 279–95.

45 Interview with a Miniaturk public relations representative and model-maker, Nov. 1, 2005.

46 My interview, Nov. 4, 2005.

47 Özdemir, *Faaliyet Raporu 2002*, 17.

48 Jean Baudrillard, "The Precession of Simulacra," in *Simulacra and Simulation*, trans. Sheila Glaser (Ann Arbor, MI: University of Michigan Press, 1994), 12.

Bibliography

Anagnost, Ann. *National Past-Times: Narrative, Representation, and Power in Modern China*. Durham, NC: Duke University Press, 1997.

Anderson, Benedict. *Imagined Communities: Reflections on the Origin and Spread of Nationalism*. London: Verso, 1991.

"Batman'a 'Miniaturk' yapılacak." *Arkitera.com*, July 21, 2003. Accessed May 27, 2017, http://v3.arkitera.com/v1/haberler/2003/07/21/batman.htm.

Baudrillard, Jean. *Simulacra and Simulation*. Translated by Sheila Glaser, Ann Arbor, MI: University of Michigan Press, 1994.

Bennett, Tony. *The Birth of the Museum: History, Theory, Politics*. London: Routledge, 1995.

Bezmez, Dikmen. "The Politics of Urban Waterfront Regeneration: The Case of Haliç (the Golden Horn), Istanbul." *International Journal of Urban and Regional Research* 32, no. 4 (2008): 815–40.

Brink-Danan, Marcy. "Counting as European: Jews and the Politics of Presence in Istanbul." In *Orienting Istanbul: Cultural Capital of Europe?*, edited by Deniz Göktürk, Levent Sosyal, and Ipek Türeli, 279–95. New York; London: Routledge, 2010.

Burke, Edmund. *A Philosophical Enquiry into the Sublime and Beautiful (1757)*, edited by James T. Boulton. London: Routledge Classics, 2008.

Çelik, Zeynep. *Displaying the Orient: Architecture of Islam at Nineteenth-Century World's Fairs*. Berkeley, CA: University of California Press, 1992.

Çınar, Alev. "National History as a Contested Site: The Conquest of Istanbul and Islamist Negotiations of the Nation." *Comparative Studies in Society and History* 43, no. 2 (2001): 364–91.

—. *Modernity, Islam, and Secularism in Turkey: Bodies, Places, and Time*. Minneapolis, MN: University of Minnesota Press, 2005.

Cizre, Ümit, and Menderes Çınar. "Turkey 2002: Kemalism, Islamism, and Politics in the Light of the February 28 Process." *The South Atlantic Quarterly* 102, no. 2/3 (2003): 309–32.

Ekinci, Oktay. "Dünya Mirasında 'Miniaturk'! … " *Cumhuriyet*, June 15, 2003.

—. "Dünya Mirasında 'Miniatürk'!" In *İstanbul'un "İslambol" On Yılı*. Istanbul: Anahtar Yayınları, 2004.

Errington, Shelly. *The Death of Authentic Primitive Art and Other Tales of Progress*. Berkeley, CA: University of California Press, 1998.

Filkins, Dexter. "Turkey's Jailed Journalists." *The New Yorker*, March 2, 2012. Accessed May 27, 2017. http://www.newyorker.com/news/daily-comment/turkeys-jailed-journalists.

—. "The Deep State." *The New Yorker*, March 12, 2012. Accessed May 27, 2017, www.newyorker.com/magazine/2012/03/12/the-deep-state.

Gibbons, Fiachra. "Turkey's Enlightenment Languishes, Like the Journalists in its Prisons," *Guardian*, March 13, 2012.

Gottdiener, Mark. *The Theming of America: Dreams, Media Fantasies, and Themed Environments*. Boulder, CO: Westview Press, 2001.

"Haliç adalarına İstanbul maketi," *Cumhuriyet*, December 20, 1989.

"Haliç'in temizlenmesi icin proje hazırlanıyor." *Yeni İstanbul*, January 30, 1957.

Hart, Kimberly. "Images and Aftermaths: The Use and Contextualization of Atatürk Imagery in Political Debates in Turkey." *Political and Legal Anthropology* 22, no. 1 (1999): 66–84.

Hitchcock, Michael. "Tourism, Taman Mini, and National Identity." *Indonesia and the Malay World* 26, no. 75 (1998): 124–35.

Houston, Christopher. *Islam, Kurds and the Turkish Nation State*. Oxford: Berg, 2001.

—. "The Brewing of Islamist Modernity: Tea Gardens and Public Space in Istanbul." *Theory, Culture and Society* 18, no. 6 (2001): 77–97.

Kadıoğlu, Ayşe. "Citizenship and Individuation in Turkey: The Triumph of Will over Reason." *Cemoti*, no. 26 (1998). Accessed on January 29, 2017. http://cemoti.revues.org/34.

Kaufman, Edward N. "The Architectural Museum: From World's Fair to Restoration Village." *Assemblage* 9 (1989): 20–39.

Keyman, E. Fuat, and Ahmet İçduygu, eds. *Citizenship in a Global World: European Questions and Turkish Experiences*. London: Routledge, 2005.

Kirshenblatt-Gimblett, Barbara. *Destination Culture: Tourism, Museums, and Heritage*. Berkeley, CA: University of California Pres, 1998.

Komins, Benton Jay. "Depopulated Cosmopolitanism: The Cultures of Integration, Concealment, and Evacuation in Istanbul." *Comparative Literature Studies* 39, no. 4 (2002): 360–85.

Kusno, Abidin. *Behind the Postcolonial: Architecture, Urban Space, and Political Cultures in Indonesia*. New York: Routledge, 2000.

Meeker, Michael. "Once There Was, Once There Wasn't: National Monuments and Interpersonal Exchange." In *Rethinking Modernity and National Identity in Turkey*, edited by Sibel Bozdoğan and Reşat Kasaba, 157–91. Seattle, WA: University of Washington Press, 1997.

Mills, Amy. "Gender and Mahalle (Neighborhood) Space in Istanbul." *Gender, Place & Culture* 14, no. 3 (2007): 335–54.

"Miniaturk Excites Us." *Gezinti*, Summer 2003, 28–31.

"Minicity açıldı." *Arkitera.com*, May 31, 2004. Accessed May 27, 2017, http://v3.arkitera.com/v1/haberler/2004/05/31/minicity2.htm.

Mitchell, Timothy. *Colonising Egypt*. Berkeley, CA: University of California Press, 1991.

Morton, Patricia. *Hybrids of Modernities: Architecture and Representation at the 1931 Colonial Exposition, Paris*. Cambridge, MA: The MIT Press, 2000.

Özdemir, Cengiz. *Faaliyet Raporu* 2002. Istanbul: Kültür A.Ş., 2003.

Özyürek, Esra. *Nostalgia for the Modern: State Secularism and Everyday Politics in Turkey*. Durham, NC: Duke University Press, 2005.

—. "Public Memory as Political Battleground: Islamist Subversions of Republican Nostalgia." In *The Politics of Public Memory in Turkey*, edited by Esra Özyürek, 114–37. Syracuse, NY: Syracuse University Press, 2007.

Pemberton, John. "Recollections from 'Beautiful Indonesia': Somewhere Beyond the Post-Modern." *Public Culture* 6, no. 2 (1994): 241–62.

Sandberg, Mark. *Living Pictures, Missing Persons: Mannequins, Museums, and Modernity*. Princeton, NJ: Princeton University Press, 2003.

"The Selection of Models." *The Showcase of Turkey, Miniaturk, The Story of How It Came to Be*. Istanbul: Kültür A.Ş., 2003.

Siegel, James T. *Fetish, Recognition, Revolution*. Princeton, NJ: Princeton University Press, 1997.

Smith, Thomas W. "Civic Nationalism and Ethnocultural Justice in Turkey." *Human Rights Quarterly* 27, no. 2 (2005): 436–70.

Stewart, Susan. *On Longing: Narratives of the Miniature, the Gigantic, the Souvenir, the Collection*. Durham, NC: Duke University Press, 1993.

Uluğ, Murat. "Miniaturk." *Yapı*, no. 262 (2003): 71–5.

—. "Miniaturk." *Arredamento Mimarlık*, no. 139 (September 2001): 38–9.

Vale, Lawrence J. *Architecture, Power, and National Identity*. New Haven, CT: Yale University Press, 1992.
White, Jenny B. *Islamist Mobilization in Turkey: A Study in Vernacular Politics*. Seattle, WA: University of Washington Press, 2002.
Zelljadt, Katja. "Presenting and Consuming the Past, Old Berlin at the Industrial Exhibition of 1896." *Journal of Urban History* 31, no. 3 (March 2005): 306–3.

6 An immersive view

In 2009, the Istanbul Metropolitan Municipality's joint stock company Istanbul Culture and Arts Products Trade Co. (Kültür A.Ş.) added to the panoply of its attractions a new "cultural showcase," entitled "Panorama 1453 History Museum." Located next to the western land walls of the historic peninsula in a municipal park with multiple attractions, this exhibitionary site boasts Turkey's first war panorama. The building has a 360-degree immersive painted view of the siege and principal battle that led to the fall of the Byzantine city to the Ottoman army in 1453 and is depicted from the perspective of the invading Ottoman army, from a point roughly where the building housing the painting is located. Curiously, the full name of the attraction drops off "conquest" but takes on "History Museum."

Despite its citizens' love of panoramic views, Istanbul never had a Panorama of its own. In not only Istanbul but also internationally, panoramas made a comeback at the turn of the millennium. The word "panorama" combines "pan" (all) and "orama" (view) and the term has two meanings: It refers to a type of perspective that "pans" or surveys the surroundings, sometimes from an elevated viewpoint, without obstructions in the foreground and with a viewing angle that requires turning the head or moving the camera to capture the view on its horizontal axis. When the first letter is capitalized, I am referring instead to the second meaning, the once-patented architectural invention. The Panorama is at the same time a spatial "medium" in which different pictorial contents, panoramas, are exhibited and messages delivered. It seeks to create the illusion of immersion in another time, location, or point of view. Panorama paintings exhibited in Panoramas conventionally depicted a historically significant event—for example, a war that took place in the past, a distant exotic place, or an elevated view from which it would be possible for the visitors see their city in its totality, a view that would not normally be physically or financially accessible on a regular basis. Thus, the Panorama was an early form of modern mass media and a form of entertainment preceding cinema. The return of the Panorama as a cultural attraction, as an exhibitionary site in an increasingly digital age, warrants closer consideration.

The popularity of Panorama 1453 is comparable to that of Miniaturk (discussed in Chapter 5), with which it is cross-promoted. However, pictured as a civilizational battle, this visualization of urban history repeats the nationalist narrative of the "conquest" and stands in stark contrast to the message of multiculturalism presented in that earlier attraction by Kültür A.Ş. Why use this "outdated" medium to deliver what many consider to be an outdated message? How is the popularity of Panorama 1453 to be interpreted? The change of focus parallels a shift in the political rhetoric of Turkey's ruling party, the AKP, since its first election to the central government in 2002, as it has sought to bolster electoral support and limited basic freedom of speech.

Based upon on-site observation, interviews, news coverage, close readings of promotional materials, and secondary literature on panoramas as well as on the urban history of Istanbul, this chapter discusses the physical, historical and political context of Panorama 1453. In addition, it describes and assesses the design of this attraction in light of precedents and theorizations of the medium, and concludes with a discussion of its political symbolism in contemporary Turkey. The existence of Panorama 1453 may be known to those who follow contemporary issues in Turkey, but a close reading of its production process, broader physical context, and connection to other attractions provides a unique perspective of this small-scale building as a governmental and cultural phenomena.

The political background of conquest commemoration

The opening statements of (former) President Gül and (then) Prime Minister Erdoğan were clear about the site's nationalist educational (or propagandistic) agenda. Erdoğan said:

> Our children will look at the future through the grandeur of that history and will say "Good For Me!" We do not want a new generation that is raised to feel inferior. On the contrary, we want a youth with self-confidence. That will happen through this [panorama].[1]

According to Gül: "this museum employs an art form that plays an extraordinary role in the formation of a historical consciousness and in reminding every member of our great nation of those days of pride in our glorious history."[2] What the AKP leadership sees as the Panorama's potential for inculcating national self-confidence is certainly one of the effects of the medium of the Panorama. Its branding as a "History Museum" is unusual and serves to place emphasis on the attraction as a pedagogic institution.

The history museum in particular orders objects in relationship to each other in a spatialized evolutionary narrative of progress in order to make visible a rendition of the past that is otherwise invisible. According to Bennett, it serves to organize and address the populace as a people (equal citizens of a nation) but also to differentiate populations.[3] Museums are theoretically open to all but culturally accessible to educated groups. Labeling a Panorama, a mass medium, as a "history museum" may be justified based on informative wall panels in the subterranean hallways leading to the Panorama platform, but on a pedagogic level, it is a deliberate political gesture. For many of its visitors, bused to the Panorama by their AKP-controlled municipalities free of charge, this may be the only instance of accessing a "museum," as well as the only performance in which they are symbolically participating in the "re-conquest of the city by the commodity and capital."[4]

Panorama 1453 has been utilized in the political machinations of the AKP in tandem with contemporary political events. The first time it was opened (it was opened twice) was two days after the Davos meeting at which Erdoğan, in support of the Palestinians, chastised Israeli President Shimon Peres by walking off a stage they were supposed to share. Erdoğan's act risked the "mediator" role Turkey had recently donned as a leader of the region, and it contradicted the "zero problems with neighbors" policy (former) Turkish Foreign Minister Ahmet Davutoğlu had been advocating until that point.[5] However, it garnered support, especially at home. At the subsequent opening of the Panorama, which he attended with his wife, an audience member (featured on the promotional DVD) carried a homemade placard that said, "The World is Proud of You."

Over its first two years of operation, (self-reportedly) 1.7 million people have visited the Panorama. It has been promoted as one of the AKP's many conquest commemoration projects (as introduced in Chapter 5). In regards to conquest commemoration, Tanıl Bora observed in 1999,

> in the popular historical narrative of both political Islam and Turkish nationalism, Istanbul is promised land .... Islamic ideology stipulates that Istanbul needs to be conquered again, owing to its degenerate state. In both Islamic and conservative-nationalist literature since the 1950s, the reconquest of Istanbul has been a popular theme.[6]

Today, the conquest of the city is still the key narrative in accounts of national greatness that legitimize not only right-wing nationalists and self-identified conservatives within Turkey but Turkey as a leader for the Muslim world. Moreover, the conquest narrative is used as a means of legitimization for the integration of Istanbul into global markets.[7]

By no means are conquest commemorations novel. Architectural historian Çiğdem Kafescioğlu's look at the immediate decades following the takeover of the city by the Ottomans in 1453 reveals that the process of commemoration had started early on, even during the siege of the city.[8] The well-known pilgrimage site of Eyüp was chosen as the location of the tomb of a famous figure, Abu Ayyub al-Ansari, a warrior during the first siege of the city by the Arabs, to legitimize the decision to inhabit the city.[9] This was an early case of "invented tradition" as well as "manufactured heritage."[10] Eyüp became the first predominantly Muslim settlement of the city, its external status to the (Christian) Byzantine city being fundamental to its choice as the place of the holy grave. It is interesting that the Panorama museum is also at the edge of the old city, in land that has thus far escaped speculative desires.

Kafescioğlu explains that it was in the so-called decline years of the Ottoman Empire that the ruling elite reconceptualized the conquest as the "golden age" and began to refer to Sultan Mehmed II simply as the "conqueror" (Fatih).[11] The capture of the city had indeed been an important turning point in the fate of the

Ottomans, which emerged as a tiny frontier principality bordering the Byzantine Empire. From early on, the Ottomans were in touch with and selectively adopted the imperial legacy of Eastern Rome. The conquest marked the beginning of the growth of the Ottomans into an Empire, mainly around the Mediterranean. Yet, its reconceptualization as a golden age came only toward the end of empire, at a point when this territory was shrinking rapidly and would culminate in present-day Turkey following the establishment of the Republic in 1923. Notably, during the early decades of the Republic, the conquest was not commemorated.

The 500-year mark in 1953 became perfect timing for elite Turkish intellectuals critical of the Republic's modernization program for obliterating ties with Ottoman civilization.[12] Perhaps the most famous of this camp is the influential poet Yahya Kemal (Beyatlı). Especially in his famous lecture, "Turkish Istanbul" (delivered for the first time in 1942 and published in 1964), he argued for the Turkish character of the Ottoman city; advocated the artistic survey and appreciation of the city; and suggested that the remembrance of the conquest was not a matter of looking backward but, on the contrary, a matter of looking forward.[13] Inspired by Kemal and his literary circle, architect Sedat Çetintaş (who preferred monumental classical Ottoman buildings over the vernacular his contemporary Sedad Eldem studied, as briefly discussed in Chapter 4 on exhibitions of old houses and streets) helped found a Monuments Commission in 1939 to design a program of celebrations to commemorate the half-millennium mark.[14] In 1950, an association (Istanbul Fetih Cemiyeti or Derneği) was set up officially to organize anniversary commemorations on May 29, starting with the 500-year one that would take place in 1953.[15] While the 500th was a government-supported event, Turkey's President and Prime Minister at the time abstained from attending the festivities because, according to the general interpretation today, Turkey had just entered NATO and wished to keep its relations with its Western allies. Since then, the commemoration festivities have been repeated on different scales but with minimal government participation or representation.

As introduced in Chapter 5 on Miniaturk, Islamism-inspired civil society groups and political parties have demonstrated a need to demote the official days commemorated by Republican history and in their place establish alternative days commemorating important events from Turkey's Ottoman past, such as the conquest of the city. Until the Panorama, such commemoration took the form of reenactments, political rallies, and project dedications. The 1453 Panorama is the first permanent simulation of the commemorated event. By revisiting this battle in its content, of course, in a highly simulated manner, Istanbul's new panorama seems to suggest a new, present-day "golden age" where, at least in spheres of the political and cultural, Turkey still commands influence on territories it once administered under the Ottoman Empire.

The nationalist triumphalism portrayed in the conquest Panorama spawned other media productions. The year the Panorama opened, the production of a feature film *Conquest 1453* (*Fetih 1453*, in Turkish) began. This was not the first Turkish film that dealt with Istanbul's conquest; the first such film was directed by Aydın Arakon in 1951 as a high budget production and with state support in preparation for the five-hundred-year celebrations of the conquest in 1953.[16] Fifty-nine years later, in 2012, with no such special date to mark, the reportedly $18.2 million budget 160-minute action film *Conquest 1453*, directed by Faruk Aksoy, opened in cinema theatres and provided renewed media attention to the 1453 Panorama History Museum.

What the new conquest film flaunts in expenditure, it lacks in artistic expression; however, this has not lessened audiences' endorsements, and it became the highest grossing film in Turkey's history, reportedly bringing in 61.2 million USD. The film is a fictionalized account from the perspective of the twenty-one-year-old Ottoman Sultan Mehmed II, who developed and directed the military strategy, including the astounding move of having the army's naval vessels land in the Golden Horn estuary. Importantly, the film opens in the holy city of Medina during the time of the prophet Mohammed, showing Mohammad's friend Abu Ayyub al-Ansari foretelling his disciples about the conquest of Constantinople.[17]

The reference to the beginnings of Islam helps to cast the conquest as a holy event, as well as a military feat, and legitimizes the idea of Turks as a nomadic conquering nation. As both journalists and scholars have noted, as the audiences of the Panorama reenact the conquest on an everyday basis, the city itself is being conquered by a real estate boom. Projects are branded with the conquest (e.g. "Maslak 1453"), demonstrating yet another instance of what Cihan Tuğal observed as the "market-oriented turn in the Islamization of the city" (in relation to e.g. Ramadan feasts).[18]

An "old" mass medium: the Panorama

As a public spectacle, panoramas have a long history, and from Walter Benjamin to Roland Barthes, cultural critics have invoked them in relation to the modern city.[19] Nevertheless, there are somewhat conflicting views on the work of the panorama. For example, writers often discuss the panoramic view and the medium of panoramic painting in relation to one another. But what does the Panorama, the modern medium-as-building featuring a 360-degree immersive painting in an indoor setting, which is also called a panorama, have to offer to the experience and understanding of the modern city?

The popular appeal of panoramic views existed long before the invention of panoramic painting at the turn of the nineteenth century and the subsequent coining of the term "panorama." However, the particular power of the medium of panoramic painting was related to the fact that, as with other techniques of simulation, it was "[relatively efficient] at providing an illusory reproduction or simulation of the real, regardless of what was being shown."[20] It has also been praised by authors from Dickens to Ruskin for its educational potential, and it is still used by political leaders for propaganda purposes.[21] Because of these qualities, it is possible to talk about the affective, emotive, intellectual, and ideological effects of a panorama—effects that work together to create its particular experience.

The panoramic view is first experienced at the surface of the skin. It is titillating to be able to survey a landscape from an elevated position. Its first quality, then, is that it is affective. According to Brian Massumi, the affective is "marked by a gap between content and effect."[22] Thus, the experience of panoramic vision lacks, at one level, subjective emotional content. It is this very quality that creates the cross-cultural, timeless appeal of panoramic views.[23]

When the Panorama was patented by Robert Barker in 1787, it was a novel spatial and visual experience. Unlike the Renaissance or classical perspective, this new medium-space provided a modern form of vision that encouraged observers to move, turn their bodies, and tilt their heads. Together with the three-dimensional objects surrounding the viewing platform, the spatial setting of the image (i.e. the lack of reference points, such as frames or a museum floor, to gauge the distance from the painting to the eye) enforced an illusion of reality. Furthermore, in city panoramas, for instance, the proximity of rooftops resulted in a haptic experience, reinforcing the "reality effect."

The second quality of panoramic painting is the emotive. It is not a coincidence that the first panoramas were of the cities in which they were exhibited. Only in the panorama could one see the city in its totality, "the city in a bottle."[24] According to Bernard Comment and Jonathan Crary, the panorama played a special role in the expanding industrial metropolis. The boundaries of this city had become less coherent; the experience of it had become more fragmented; and in it the individual felt increasingly isolated.[25] Comment writes, "The reaction to this general loss of 'readability' in and of urban space was two fold: *induction* and *panoramism*."[26] One could no longer experience the whole city or get a sense of its boundaries, size, and details—except in the panorama. In relation to Istanbul, sociologist Ayşe Öncü suggests the panorama "encapsulates a *feeling* we can never retrospectively imagine, a sense of wholeness with the city."[27]

The third quality of the panoramic view and the painted panorama is that they bring together not only here and there but also past and present. Roland Barthes suggested the panorama of Paris as seen from the Eiffel Tower put the observer to work—the work of "decipherment" (deciphering). This, he claimed, is the "intellectual character of the panoramic vision."[28] The panorama thus denies the observer mere aesthetic appreciation: "To perceive Paris from above is infallibly to imagine a history."[29] In other words, regardless of whether one is looking at a contemporary urban scene, one must keep on deciphering, structuring, and filling in the historic quality of the scene in one's mind.

Finally, it is common to interpret the Panorama and the panoramic view in relation to Michael Foucault's account of Jeremy Bentham's model prison, the Panopticon. This was invented at around the same time as the Panorama, and the two together marked the beginning of the modern era.[30] In this regard, Tony Bennett and Jonathan Crary have pointed out that despite Foucault's "dismissal" of the spectacle, the societies of discipline (as described by Foucault) and of spectacle (as described by Guy Debord) overlapped, and were actually inversions of each other. Mitchell and Bennett have likewise discussed how the medium of the Panorama was central to the imperialisms and emergent nationalisms of the nineteenth century. It presented "the world as a picture" for domination, allowing viewers to identify with the eye of power.

The "panoramic" thus became a dominant mode of seeing in the nineteenth century, manifest not only in the buildings that exhibited painted panoramas but also at world exhibitions, natural history museums, and similar venues of spectacle that surveyed, categorized, ordered, and displayed the world. These, too, put the viewer at their center as both observer and observed, thus appealing to the desire to see and be seen.

Panoramic urbanism in Istanbul

Istanbul's unique topography, planning, and architecture create effects that I describe as "panoramic urbanism": Panoramic views are not only desired but are also readily available as a source of pleasure and identification with the city. Panoramic urbanism as a "way of life" is characteristic of the city—rendering unnecessary or unappealing any Panorama (as building) until the opening of Panorama 1453 in 2009.

Istanbul is a city designed by the Ottomans to be experienced panoramically, but it was European visitors who produced its first panoramic pictures. Urban historian Ekrem Işın suggests, "… the medieval [pre-Ottoman, Byzantine] city [did] not present [such] a panoramic image." Indeed, the earliest map of the city, Cristoforo Buondelmonti's bird's-eye sketch dated 1422, offers a view of the city constrained within medieval walls, clearly demarcated from its surroundings.[31] It was not until the "Vavassore map," printed in the 1530s, after the Ottoman takeover, that the city's walls lost their prominence vis-à-vis depictions of the urban fabric and the waterways, the Bosphorus and the Golden Horn, were now shown as active and busy with vessels.[32] But then, with Melchior Lorich's 1559 drawing, the depiction of the city switched entirely. A twelve-meter panorama with several focal points, it paralleled a shift in thinking about the city to emphasize its horizontal expansion and new skyline of minarets and domes.[33]

Panoramic views of the city were produced long before their utilization in the context of nineteenth-century imperialism in various mass attractions and vision devices. Yet, even then, what the producers intended in popular displays throughout the nineteenth-century century never fully corresponded with the diversity of meanings and understandings brought to them by spectators. For instance, according to Denis Blake Oleksijczuk's study of the first panoramas, the medium of the Panorama did not necessarily offer an empowering experience that reinforced identification with the imperial state; many visitors found them destabilizing and overwhelming.[34] Therefore, it is important to situate the subjects and contexts of viewing.

The Ottoman imperial city was planned with monuments on commanding hilltop locations. It was this planning regime that gave Ottoman Istanbul its unique skyline. The most pointed example of how the city came to be experienced panoramically under the Ottomans is the siting of the Topkapı Palace on the eastern tip of the historic peninsula. In the middle of the palace, the Tower of Justice overlooks the surrounding landscape, its window once symbolizing the personal presence of the sultan among his subjects.[35] The view from above, however, has never solely been in the domain of the sovereign or the privileged. Across the city, hilltops have offered common people viewpoints onto its waterways. This topography still connects the dispersed city, with its ever-expanding outskirts—if only in visual and imaginary terms. In environmental design terminology, it provides a "mental map" as well as "sense of place."

The desire to reach out for views of the water, using not only topography and siting but also architectural design, has a long history along the Bosphorus.[36] The elevated views afforded by the city's many hills and its unique waterside architecture are truly a source of communal enjoyment. In recent years, rooftop restaurants and bars have elevated this form of scopic entertainment to what may be called "panora-mania."[37] One of the city's trendiest restaurants in the past decade, 360° Istanbul, flaunts this attitude in its name, branding, architectural design, and even furniture detailing. In an effort to theorize this desire for panoramic views, political scientist Engin Işın observes, "Every building, it seems, tries … to orient itself to the Bosphorus in order to catch a glimpse of its glorious glisten and glitter." Işın has proposed the word *keyif*, roughly translatable as "enjoyment," to describe the pleasure afforded by the views of the city.[38] As a defining mood, one that binds Istanbul's citizens and is specific to it, it stands as a counterweight to the novelist Orhan Pamuk's more well-known sense of *hüzün*/melancholia.

In social theory, elevated views are generally associated with a desire to control and possess terrain—to create a distance between oneself and the world—and as such, they have been linked to Western imperialism. Especially relevant is political scientist Timothy Mitchell's description in his classic article "The World as Exhibition" of attempts by Europeans to "set up the East as a picture." They did this, for instance, by establishing commanding views from elevated positions such as the Great Pyramid of Giza, which they turned into a viewing platform. For example (as the story is retold by Mitchell), on his first day in Cairo, Gerard de Nerval meets a French photographer-painter with whom he looks for a suitable point of view, to find it only outside the city.[39] In a similar vein, Michel de Certeau—in another classic text, "Walking in the City"—builds on Michael Foucault's discussion of the Panopticon to suggest that the elevated view belongs to the logic of top-down power, the corporation, the institution, and the state.[40] As a form of resistance, down in the street, pedestrians can use the given terrain tactically and imaginatively, their behavior never fully determined by imposed order.

Scholars of Istanbul have noted the increase in the numbers of corporate and luxury high-rises in the past decade and their branding in relationship to their views: Ayşe Öncü highlights the fact that there is an integral relationship between the privilege of views and socioeconomic status in the current residential make-up of the city.[41] Even before the current era of neoliberal urbanization, the housing stock of the city, mainly characterized by walk-up apartment buildings, displayed vertical social segregation with upper level units with views having prime status.[42] Specific to Istanbul, however, elevated views are neither restricted to buildings of privilege nor solely available from points outside the city proper. One has ready access to panoramic views from within the city, by merely standing on an elevated location, a hilltop, or on the banks of the Bosphorus. Such views offer residents the possibility of immersion in the city without having to travel outside of it. Not all views from above are totalizing; to assert so would be to totalize the function of the overall view. Furthermore, there is a difference between a totalizing aerial view and an elevated view. In Istanbul, the scopic drive to orient oneself in the city and the pleasure derived from it likewise needs to be differentiated from demands for "legibility" and control.

The site and context of Panorama 1453

Panorama 1453 is located in a municipal park in Topkapı, outside the Byzantine city walls (see Figure 6.1). It is bordered to the northwest by the E5/D100 highway and to the northeast by Topkapı Cemetery. To the east, inside the city walls, is a residential area of mainly concrete-frame apartment buildings, and to the west is an industrial area with a number of production plants (see Figure 6.2).[43] Close by is Sulukule, the historically Roma neighborhood which was expropriated and demolished in the second half of the 2000s to make way for a new "Ottoman"-looking residential neighborhood of real estate offered on the private market.

This area, right outside the historic city walls, was formerly a somewhat chaotic transport interchange with an intercity bus terminal, numerous bus stops, and many peddlers. As such, it served as an entry point to the city for travelers and rural-to-urban migrants and was typically jammed with traffic. However, in 1999 the Istanbul Metropolitan Municipality decided to move the bus terminal further west to Esenler, and its Directorate of Projects carried out, during the mayoral term of Recep Tayyip Erdoğan, one of the largest "urban transformation" projects in the city's history—in size, at least, "comparable to Hyde Park" at thirty-five hectares.[44] "Urban transformation" (in Turkish, *kentsel dönüşüm*) is a term favored by local officials, and now used by academics, to refer to the opening to investment of potentially profitable spaces in accordance with neoliberal economic policies. What the term neglects to acknowledge, of course, is the displacement and dispossession it brings for existing communities and the inequalities it creates by selecting specific actors to profit from it.

This project was at first dubbed "Topkapı City Park," but it was renamed "Topkapı Culture Park" to reflect the choice of Istanbul as a 2010 European Capital of Culture.[45] The scheme, as implemented, used eminent domain—which ought to be ordinarily used for the public good rather than for profit-making purposes—to demolish properties on and abutting the site. It then reorganized surrounding and through roads and public transportation, and a new minibus terminal was constructed, as well as offices to be rented out by the municipality. The open-air park itself includes an amphitheater, a "sosyal tesis" (social facility—a combined teahouse/café and restaurant), a "culture village," a "traffic education park" (which is described as a "miniature city" targeting nine- to fourteen-year-olds), and green areas for walking. With the relocation of the bus terminal, peddlers were also banished.[46]

Although this project opened up space for public use and streamlined transportation links, Topkapı did not necessarily become more accessible in the sense

Figure 6.1 Panorama building in context.

Courtesy of Kültür A.Ş.

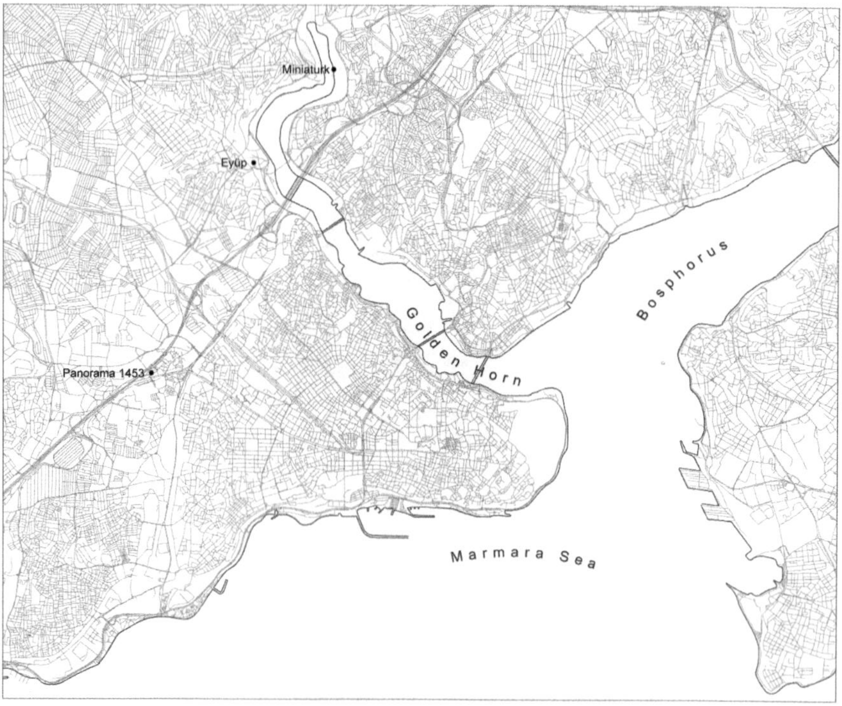

Figure 6.2 Panorama building's location in the city.

of sociability as a result. The new arrangement of public transportation includes a stop for the metrobus (a fast bus service with its own lane) nearby, and a tram stop is located right behind Panorama 1453. Hence, the site is extremely accessible via public transport. Yet, because it adjoins a residential area on one side and an industrial one on the other, and because a major highway runs through it, the site-as-park does not bring people together, and it remains underused (see Figure 6.3). Visitors arrive mostly to visit Panorama 1453 and depart, without wandering around.

According to the district municipality's website, the eventual goal of the urban restructuring effort is to extend the park in subsequent stages along the old city walls to create a "cultural island" that supports religious tourism.[47] There are three mosques in proximity. One, the sixteenth-century Takkeci (Takyeci) İbrahim Ağa Cami, was restored during the creation of the park. Several old cemeteries also border the park, and nearby are the monumental tombs of past prime ministers Adnan Menderes (1950–60) and Turgut Özal (1983–91), both known for their economically liberal, conservative populism, and during whose terms the city went through dramatic transformations. Ultimately, however, the site's religious appeal derives from its historic importance as the location of the principal battle that led to the fall of Constantinople during its siege by the Ottoman Turks—the

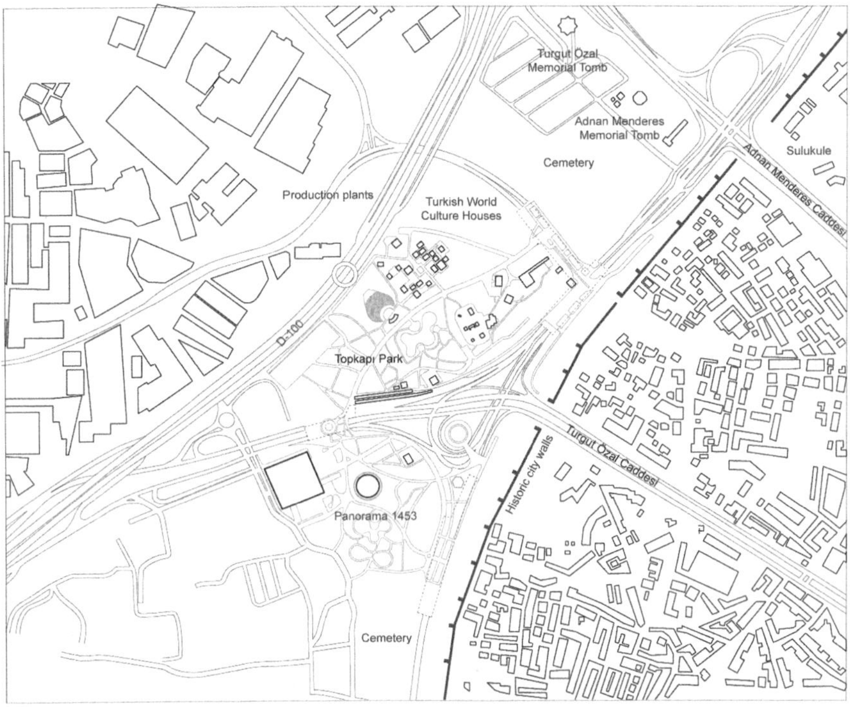

Figure 6.3 Panorama building's location in its immediate urban context.

subject of the panorama painting inside the Panorama History Museum. In addition, the site is relatively close to Eyüp, an important pilgrimage site for Muslims.

Several attempts have been made to enhance the park's appeal through additional programming meant to foster ethnic nationalism. Principally, this involves an installation now called "Turkish World Culture Houses"—a local version of Disney's Epcot World Showcase without an organizing central pond. The compound focuses on the Turkic world and is composed of small wooden buildings displaying features meant to resemble the "Ottoman House." There is also a yurt (a traditional circular tent) and an ethnic restaurant, run by a Uighur family from western China, where Turkish visitors can taste the delicacies of the kitchens of "old Turks" still living in Central Asia.[48] In 2003, the compound was first designed to be an "Ottoman Village," where traditional crafts would be displayed.[49] At some point, however, its conceptual scheme changed, and each house is now dedicated to a Central Asian or Turkik country—Azerbaijan, Kazakhstan, Kyrgyzstan, Turkmenistan, Uzbekistan, Tataristan, Northern Cyprus, the Balkans, and the Turkish Republic of Northern Cyprus. The embassies of these countries have been invited to exhibit ethnographic artifacts representing their cultures, and to hold their national celebrations there. Additionally, the Culture Village has been designated as the official celebration venue for the annual Nevruz festival.[50]

The origin of this cultural attraction is tied to world events. Following the dissolution of the Soviet Union, Turkey reached out to the newly independent countries of the Caucasus and Central Asia; it recognized their sovereignty before anyone else, established embassies, and began forging cultural and economic ties. Meanwhile, back in Turkey, the government began hosting summits on "Brotherhood and Cooperation of Turkish States and Turkish Speaking-Communities."[51] These summits were halted in 2001, but they were revived by the AKP in 2006, at which time Prime Minister Erdoğan made a proposal to establish a United Nations of Turkish-Speaking Countries. While Turkey has newfound influence in the region through the missionary schools of the religious sect of Fethullah Güven (which had links to the AKP until recently), the AKP's motivation at home was to appeal to the electoral base of the Nationalist Movement Party (Milliyetçi Hareket Partisi, MHP). While Turkey never acquired the political leadership role it envisaged for itself as the region's "window onto the world," Turkish companies, especially construction companies, have benefited immensely from ties to these new states, helping build their new economies. In recent years, more visibly than ever, citizens of these countries have joined the informal labor pool in Turkey.

The designated countries in the "Turkish World" compound have not visibly invested in the houses in the park thus far. However, these little structures may well function in the near future as meeting places for their citizens. Regardless, the Panorama, which celebrates "Muslim Turks" conquering Christian territory, becomes additionally symbolic when sited next to such a compound that winks toward Muslim-Turkik "brotherhood." More importantly, the changing orientation in Turkey's foreign policy under the AKP and shifts in the AKP's internal electoral base relate to the choice of a Panorama, an "old" mass medium, featuring an overfamiliar rendition of a particular historical moment, to emphasize a specific approach to nationalist education.

The design of Panorama 1453

On a typical Sunday afternoon in 2011, the wait to get in Panorama 1453 could last up to an hour, with lines stretching all the way to the old city walls nearby. According to figures provided by the management, on average, 3,000 people—mostly families during the summer and schoolchildren during the winter—visited it per day, and up to 9,000 could arrive on weekend days.[52] Of this average, I was informed, about 400 were tourists, mainly from Arab countries. The management works closely with tour operators, journalists, news agencies, and schools to boost attendance. The operational chief reports that, on average, fifty news reports appear on Panorama 1453 each month, with ten to fifteen of them on TV. A televised discussion show on Middle Eastern politics was even hosted inside Panorama 1453.[53] Far-away municipalities of Istanbul districts sponsor free mass tours to the attraction for their residents. The management invites consulates to Panorama 1453 regularly for receptions. International guests of the municipality and the government are also brought to there. In addition, a free shuttle operates between the religious shrine of Eyüp and Miniaturk, across the Golden

Horn. Finally, these sites cross-promote each other—for example, entry tickets to Panorama 1453 advertise Miniaturk on the back—leading their managers to believe they share about seventy percent of their audiences.

From the outside, Panorama 1453 consists of three volumes: a cylindrical rotunda housing the panorama is abutted by a two-story rectangular block fronted by a small, one-story entry hall. The large cylindrical volume is topped by a conical roof, and its windowless sides are covered with vignettes of Old Istanbul, reproduced from late Ottoman-era etchings. The way these are set in arched frames is intended to simulate the experience of looking out from within a colonnade, so that walking around Panorama 1453, it will seem to visitors that they are looking out to Old Istanbul from within the present-day city. The secondary rectangular volume is also solid, except for a row of seven domestic-scale windows facing the entrance on the second floor, which provide daylight for administrative spaces. The logo of the municipality and the name of the museum are affixed to this same façade on top of the windows, as if an afterthought.

The third, smallest volume, which I refer to as a hall because it receives visitors, is of glass, and is topped by a bronze-colored, sloped roof. It looks more like it was intended to protect visitors from wind than to serve as a functional space, but the ticket booth is tucked between its in and out doors. A security guard waits there at all times. Once past this tiny space, one enters a formal foyer, with an elevator and stairwell to the right and a gift shop to the left where one can purchase issues of Kültür A.Ş.'s "culture and arts journal" entitled *1453*, among other memorabilia. The entry sequence invites the visitor up the stairs located next to the entrance door, but the flight going up is cordoned off—upstairs areas are occupied by offices and a room for VIPs. Instead, visitors are guided downstairs. Descending the dark stairwell, with *fer forgé* balustrades and a side wall covered by a mural sculpture of the conquest battle, one arrives at the first basement level, housing a second foyer with wall exhibits and restrooms.

Descending further to the second basement, one enters a long corridor leading to a narrow, dark stair at the end that ascends to the viewing platform. This second basement level and its corridor and foyer space again contain wall exhibits. These offer a combination of images of historic documents and an abundance of explanatory text, all arranged in graphic designs dominated by red-yellow hues that seek to reproduce the mood of royal-military grandeur. The animation of a war scene is also displayed here on one of the walls, using an LCD screen. The exhibited plates have titles such as "A Miracle of the Prophet: The Conquest," "Constantinople Ravaged by the Crusaders," "Ottomans from the Establishment to the Conquest," "Why Was the Conquest Necessary?" "Ottoman Fleet in Front of Constantinople," "Cannons of the Conquest," "The Breaking Point of the Siege," "A Turning Point of the Siege," "Preparations for the Final Attack," "Final Address in Front of the Walls of Constantinople," and "The Prayer of Sultan Mehmed II." These provide a chronological narrative that progresses as the viewer proceeds through the waiting line toward the platform. The rest of the exhibition in the circulation and foyer spaces concentrates on the cult personality of "the Conqueror." The plates are reproduced in the official book of the Panorama, and English translations to titles are provided. But the fact that only Turkish text is available in the exhibition space suggests the Panorama is intended mainly for a local audience. Lastly, it is a narrative solely of the "winner" and is overtly nationalist in tone.

The study models visible just before one arrives at the stair to the viewing platform show how a semi-spherical volume covers the platform instead of a cylindrical one (see Figure 6.4). Their domes are opened up on one side so that the visitors can have a bird's-eye view and a clear understanding of how this Panorama works. These study models show how the horizontal surface between the platform and the painted wall has been elaborated as faux terrain covered by *attrapés* (hoax objects). The visitors are ushered up the stairs onto the platform by a Janissary-costume-wearing attendant. There, they find themselves not only viewing a picture but in the middle of a theatrical stage that seeks to transport the viewer into the middle of a field of war via a multimedia presentation. At least, this is the intention, but this is only partially achieved due to the constraints of the Panorama as a medium.

Overall, Panorama 1453 is insufficiently designed for its role as a "museum." The volume and the scale of the entry do not communicate its public function, and the building turns its back to the main points of public access. Entry and exit are from the same singular curricular stairwell, and there are no resting areas. Moreover, the scale of the building is not suited to the large audience it receives. This is not surprising considering the haphazard way the project was commissioned and realized. After interviews with the artistic director and the operational chief, I was not able to identify an architect at all.[54] According to the artistic director, the Istanbul Metropolitan Municipality's Directorate of Projects was responsible for the building, and the dimensions of the Mesdag Panorama in The Hague were simply copied to create the volume housing the painting. The rest was improvised.

The artistic director himself had been in the animated-film business prior to taking over responsibility for creating the panorama. He offered his services after seeing

Figure 6.4 Models on exhibit.

the building under construction. As a result of its success, however, plans were drafted to create other such installations. Among them, the artistic director and his team, now based in an office on the grounds of the park, were working on the 1915 Panorama for Gallipoli—an idea, he claimed, he successfully pitched to the Prime Minister at the opening ceremony of the Panorama 1453.

Both the medium-building and the painting exhibited inside are important as governmental and popular culture phenomena, rather than as artifacts with architectural or pictorial merits that surpass their nineteenth-century predecessors. However, it is still important to understand the predecessors to understand that this attraction functions as a site where urban history is visualized. This lengthy description serves to demonstrate the haphazard nature of the final design experienced by the visitors—despite this haphazardness, the spatial experience remains important to interpreting the intentions behind the medium and its potential effects on visitors.

Museums store and display "things" to inspire their visitors in certain ways. In fact, Carol Duncan argues that (art) museums are ceremonial monuments; they work like temples, to which visitors arrive with the willingness and ability to shift into a certain state of receptivity.[55] In reference to the Panorama, Griffiths suggests that the experience of ascending from the dark to the light of the platform, and the feeling of being transported elsewhere, constitutes a "quasi-religious experience."[56] The experience of the nineteenth-century Panorama falls between the cathedral and the motion picture. A visit to Panorama 1453 exists somewhere, in the Turkish context, between visiting a mosque and viewing a motion picture, perhaps. Since a dome is used in Panorama 1453 in lieu of a cylindrical enclosure wall, the mosque becomes even a better architectural analogue. Visitors arrive with varying degrees of preparedness to be transformed. However, visitors' receptivity and "transformation" always depends on their existing belief systems and subject positions as well as interactions with the exhibit and inside the exhibit with other patrons. Unlike the art museum to which Duncan refers, and as explained before, Panorama 1453 is a highly interactive space, not a silenced one, that is carefully curated for elation and identification.

The appeal of an "old" medium in the age of new media

Upon opening, Panorama 1453 did not immediately attract much critical commentary in the local press or detailed analysis in scholarship—the few exceptions astutely focused on its lack of historical veracity and questioned its ideological implications.[57] Undoubtedly, the subject matter of the panorama and its method of one-sided rendition, glorifying the conquering "Muslim Turks" (when the Ottoman army was quite heterogeneous at the time), legitimize violence in the manner of authoritarian regimes and serve present-day interests. However, no one, it seems, has reflected on the particular choice of the Panorama as a medium to present this message. I have discussed (in Chapter 5) that theme parks are proliferating, especially in the booming economies of the Middle East and Asia. For instance, Thomas Campanella writes that between 1990 and 2005 some 2,500 theme parks opened in China, and he describes how the entire consumption landscape of China is themed in one way or another.[58] Does the Panorama find a similar niche in the symbolic economy of cities? What is the appeal of an "old" media in the age of new media?

It seems strange to see new Panoramas today when it is generally assumed that cinema wiped them out as a medium at the turn of the twentieth century. But Panoramas were rediscovered, as it were, in the late 1970s and 1980s. Their rediscovery coincided with the emergence of experiments and writings on virtual reality, of visual culture studies, the rediscovery of the history of the panorama as the first "mass medium," and calls for the conservation of surviving panoramas such as the Mesdag Panorama in the Netherlands. These concerns led to the establishment of the "International Panorama Council" in 1992.[59] The 2010 conference of this organization was held in Istanbul, with the sponsorship of Kültür A.Ş. and Istanbul Metropolitan Municipality. Images and statements from this meeting were later used by Kültür A.Ş. as proof of the innovative aspect of the museum in its promotional DVD.[60]

Most new panoramas are of battles with clear nationalistic messages. However, there was far more variety in subject matter when the medium was first introduced. When the first public shows took place in London and other European cities at the turn of the nineteenth century, roughly three kinds of imagery emerged. The first group included views of cities in which the panoramas were exhibited, such as Edinburgh and London. They were generally painted from elevated locations such as towers and rooftops. The second type were panoramas of important battles.[61] These were shown most frequently at eye level so as to immerse the viewer in the scene, as is the case with Panorama 1453. China and Korea seem to have quite a few contemporary panoramas that depict such battles.[62] War was a popular topic for early panoramas, too, because such installations helped keep wars in the public eye.[63] Stephen Ottermann (and more recently and more thoroughly Denise Oleksijczuk) thus explains how the subject matter of early war panoramas—celebrating British imperialism and, in particular, the successes of its navy—were key to the public appeal of the medium. In other words, not only the technology but the subject matter is important. And, in turn, such imaginings of distant territories became fundamental to justifying imperial expansion to the general public.

The third type of panorama was of foreign cities—again from elevated locations. According to some critics, these served as replacements for actual travel.[64] Istanbul was the first foreign city whose panorama was exhibited in London in 1801.[65] The selection may be interpreted in light of the concurrent alliance between the British and Ottoman Empires against the French and Russians following the invasion of Egypt by Napoleon. However, Reinhold Schiffer presents another view: that, according to visitors, Istanbul was the most panoramic city in the world, and it simply lent itself exceptionally well to the medium. As he writes:

> British travelers' visual experience of Istanbul in its totality was intimately linked to the panorama. … The great majority of visitors agreed that Istanbul was the most panoramic city in the world. It is not surprising, therefore, that the city became an early subject for panoramic exhibitions.[66]

Interestingly, the panorama experience of Istanbul in London later informed the panoramic experience of the city on-site for European visitors. Schiffer thus suggests that subsequent visitors came to consider Istanbul as an aesthetic object, set in the landscape. Because it was a city that needed to be appreciated from a distance, they appropriated the category of the picturesque to explain their experience of it as an "Oriental" city, overlooking its commercial and political life.

As they were developed in the nineteenth century, the outer walls of a typical panorama installation were cylindrical. The interior was naturally lit, viewed from a raised platform (belvedere) covered by a canopy. However, as the study models on exhibit in Panorama 1453 clearly show, it is set within a dome. It also lacks a canopy, and thus a belvedere feeling, and is artificially illuminated. Its painting, depicting the siege of the city, is therefore hemispherical, showing a full sky with clouds, without the illusion of looking from within a belvedere enabled by the canopy. The painting was actually created by eight artists over three years, working on computers and in a nearby workshop where they built props. The artists decided to print out the painting they had created in the computer in pieces, using sheets whose edges they touched up once they were pasted to the interior surface of the dome.

The painting successfully incorporates many of the visual techniques of early panoramas. The Ottoman army constitutes the foreground, and its units partially block the view to achieve a proper immersive effect. One highlight, according to official promotional materials, is the image of the sultan on a white horse surrounded by his religious advisors. Another is the figure of Ulubatlı Hasan, a young soldier who first erected the Ottoman flag atop the Byzantine fortress. Thus, the painting presents a composite view, combining a topographical sense of the battlefield with depictions of noteworthy military figures and anecdotal moments. A voiceover further relates the story of the siege in glorifying detail and provides information on the different kinds of units that constituted the Ottoman army. Finally, a surround-sound track animates the diegetic space by fictively recreating the noises of battle. The painting itself shows that the previously unbreachable walls of Byzantium have clearly been destroyed, and in the distance a church signifies that the territory soon to be conquered is Christian space.

What effect is this scene of war supposed to have on viewers? Officials with direct connection to the AKP are not troubled by its celebratory quality. In this regard, their views mirror the AKP's and Erdoğan's problematic, exclusionary approach to non-Muslim citizens of Turkey. It seems the artists are the only ones who have problematized the scene they painted. In a promotional DVD film, the artistic coordinator uses the word "affection" to express his goal for the panorama. He says his wish was that the panorama would raise curiosity among viewers to further explore history. In an interview with me, he further acknowledged the problem created by a depiction of Islam's triumph over Christianity, of Ottomans over Byzantines, in a city with a living Rum (Greek Orthodox) population. He explained that, as artists, his team had tried to provide equal space and roles to both parties by choosing a point of view from the walls themselves. But they were not able to achieve a satisfactory technical result from this perspective and settled with a point of view from the Ottoman side.

Why is the "old medium" of the Panorama still popular with viewers? When asked, officials tended to respond by talking about its technical aspects. In particular, they highlighted Panorama 1453's success in the painting's verisimilitude, which causes dizziness for "ten seconds." The Panorama itself is located right outside the walls whose breach the painting inside depicts. To produce it, a panoramic photograph was taken at a height of 3.5 meters in front of the nearby gate in the old city walls, and this was used as a basis for the painting's topographical details. The horizon was set at a height of roughly 1.5 meters in relation to the platform—between a child and an adult's eye levels. Together, officials report, these technical factors cause some visitors to become confused and think they are actually outdoors when they ascend to the viewing platform.[67]

The manager and the artistic coordinator have emphasized how the "full-panoramic" nature of the Panorama, with a semi-spherical sky view, was their innovation and that it contributes substantially to the "reality effect." As mentioned earlier, however, the ceiling over the viewing platform (its canopy) was traditionally considered essential to the medium's illusion, while the use of artificial light in Panorama 1453 fails to create the contrast that natural light entering through a glass roof would have created in a nineteenth-century Panorama. Another quality that is lacking is the "verisimilitude of the experience," as Griffiths identifies in the nineteenth-century Panorama, "since moving clouds and variations in light create tonal shifts reminiscent of the contingent nature of outdoor viewing."[68] In essence, then, the dome and artificial lighting may lessen the reality effect. Viewing the study models prior to entering the platform also diminishes the illusionary impact.

The artistic coordinator suggests that the unprecedentedly high level of detail, made possible by the use of computer graphics, may be partially responsible for the panorama's success. The artists sought to satisfy viewers who would be zooming into details armed with high-resolution digital cameras. In ordinary perspectival painting, the farther away an object is, the less detail is provided, mimicking the inability of the eye to see into the distance. However, in this instance, providing extra detail does not necessarily help the illusion; but it does help viewers interact with the painting via their cameras, to zoom in search of details. Three-dimensional replicas of war equipment in the foreground, such as cannons, also heighten the somatic sense. Since multiple focal points are on offer, the viewer has to decide what to look at, and thus must continuously adjust her gaze. Thus, looking at the soldiers fighting, sometimes the viewer feels she is hovering, and at other times she can establish eye-to-eye contact with individual figures. In many cases, the picture features telescopic details of faraway scenes that would not normally be visible with the naked eye (see Figure 6.5).

Perhaps an analogy with theater, rather than cinema, is more appropriate. İlber Ortaylı, a respected Ottoman historian in Turkey, the former director of the Topkapı Palace, and an occasional advisor to the municipality, invokes this comparison to explain the power and choice of the panorama as a medium.[69] "Theatricality" is indeed the effect that not only the Panorama but the whole culture park, with its set-like design, seeks to achieve. In this sense, the appeal of the

Figure 6.5 The painting of the 1453 "conquest" exhibited inside Istanbul's conquest panorama overlaid with enlarged details (left to right), demarcating the space behind the walls as Christian space with churches and crosses; depicting the anecdotal moment of Ulubatlı Hasan erecting the Ottoman flag; and Ottoman Sultan pointing and his advisor next to him praying.

Manipulated from source image, courtesy of Haşim Vatandaş.

panorama derives from its invitation for "reenactment" and interaction more than immersion.

A comparison with the virtual reality show "Sky Ride Istanbul"—on top of Istanbul Sapphire, a high-rise building in the city's new Central Business District along the Levent-Maslak axis—brings out just how interactive the Panorama actually is.[70] At Sky Ride Istanbul, the audience wears 3D glasses and buckles in to fly in a virtual helicopter over Istanbul—departing from the seat location in the theater atop Istanbul's tallest building, Sapphire, and being exposed to atmospheric effects such as wind and water along the way. The virtual ride silences and demobilizes the audience; its atmospheric effects distract the rider's focus from the actual city view, turning the ride into an introspective experience about technology. By contrast, the naked-eye view on the platform of the conquest panorama allows visitors to interact with each other, move about, and, using their cameras, to zoom in and out (see Figure 6.6). The panorama affords them, not an overall view of the city, but a movement in time. Unlike the virtual ride, then, the Panorama is not a space where viewers metaphorically "lose" their bodies. On the contrary, they engage in conversation with each other, wander around, and take pictures; the platform of the panorama is an embodied and social space, irreducible to the message of the panorama painting that is exhibited (see Figure 6.7).

Visualizing urban history in the Panorama

My discussion of Panorama 1453 has sought to understand this architecture-as-medium by situating the medium and the message (the constructed idea and ideal of the conquest) in relationship to the broader physical site and its planning, the political context as well as the politics of the medium itself.

Panorama was invented at the beginning of the nineteenth century in Western European metropoles as a mass attraction, where it came to epitomize an

Figure 6.6 Visitors interacting with the painting with their cameras.

Figure 6.7 Visitors moving and interacting on the platform with each other. Janissary-costume-wearing attendant is seen resting.

international hunger for physically, geographically, and historically extended vision. However, it was not appealing for audiences in the Ottoman Empire, despite the fact that there is a long tradition of admiring, designing, and planning to enhance panoramic views of the city, and despite the fact that panoramas of Istanbul were of great appeal in Europe among European audiences.

Scholars of the medium have questioned whether producers' intentions and messages were fully effective on the audiences (though we lack "audience studies" of the medium) and whether all members of an audience would receive the messages of a war panorama in the same way, since messages are received according to the subject positions of the viewers—not to mention that the medium itself has spatial qualities that incapacitate a uniform message. My discussion highlighted the difference in intentionality between the multiple producers and contributors to the 1453 Panorama History Museum; it also described and analyzed how Panorama 1453 compares to its early nineteenth-century precedents to assess its relative efficacy in delivering its message.

Panorama 1453 is a seemingly simple yet historically layered message. The conquest has been a popular topic for conservative politics. As articulated for the 500th celebrations in 1953 by a group of intellectuals who formed the "Conquest Association," the reenactment and commemoration of the conquest in a variety of media would serve to restore "Turkish Istanbul" and to save it from overly-Westernized, secular classes. Since 1994, with Erdoğan as Istanbul's mayor, conquest commemoration served as a symbol of the growing influence and power of the conservative right against the secularist establishment. Then, from 2002 with the AKP under the leadership of Erdoğan, the idea of the conquest evolved into a celebration of Ottoman history to market Istanbul and attract more tourists and global capital. By the end of the 2000s, Turkey's leadership was trying to position the country as a leader for the Muslim world—the limitations of which effort would soon be revealed during the Arab uprisings in the early 2010s. Thus, the appearance of Panorama 1453 in the cultural landscape of the city needs to be viewed in conjunction with these changing dynamics.

Looking at a contemporary site, this analysis admittedly lacks historical distance. However, the historical analysis of open-air parks and theme parks in Chapter 5 and panoramas in this chapter suggest how both mediums responded to a hunger for extended vision and guides my interpretation of such sites in the present. In contemporary Istanbul, Panorama 1453 epitomizes the ruling party's ideological hunger for a geographically and historically extended sphere of influence for Turkey and its abandonment of the idea of multiculturalism displayed in Miniaturk for pan-Islamism, one that is inclusive of pan-Turkism. Panorama 1453's central subject matter, of heroic Muslim Turks conquering a formerly invincible Christian enemy, is referenced, despite its lack of veracity, by other forms of media, from popular feature films to TV series. Some of the visitors of Panorama 1453 are also consumers of these other kinds of media. At the same time, Panorama 1453 is situated in urban space and works as part of an ensemble of attractions in a park setting, as well as constituting one node of a symbolic triangle that connects the holy shrine in Eyüp and the open-air, nation-themed miniature park Miniaturk. It provides an anchor to a visual culture saturated with "Ottoman" labeling, from cuisine to real estate. Thus, Panorama 1453 is received in relationship to visitors' exposure to any of these, or the lack thereof.

Notes

1 My emphasis. Erdoğan: "İstanbullu hemşerilerimizin hizmetine sunmak da şimdi hamdolsun Kadir başkanımıza nasip oldu. Sayın Topbaş'ı ve bu projelerde emeği geçen mimarından, mühendisine, işçisine kadar, yüklenici firmanın değerli yöneticilerine kadar herkesi şahsım, İstanbullular, ülkem, milletim adına tebrik ediyorum kutluyorum. Yavrularımız, geleceğe o tarihin mehabetiyle bakacaklar ve 'Ben neymişim' diyecekler, bunu görecekler. Bir aşağılık kompleksi içinde yetişen gençlik istemiyoruz, tam aksine kendine öz güveniyle yetişen bir gençlik istiyoruz. İşte bununla olacak bu." Istanbul Metropolitan Municipality website, "Türkiye'nin ilk panoramik müzesi 'İstanbul'un Fethi'ni' yeniden yaşatacak," January 31, 2009, accessed on May 29, 2017, http://www.ibb.gov.tr/tr-TR/Lists/Haber/DispForm.aspx?ID=17004.

2 Abdullah Gül, 15 March 2009, Guest Book, reproduced in *Panorama 1453 Tarih Müzesi, Panorama 1453 Historical Museum* (Istanbul: İstanbul Büyük Şehir Belediyesi, 2009).

3 Tony Bennett, "The Political Rationality of the Museum," in *The Birth of the Museum: History, Theory, Politics* (London, New York: Routledge, 1995), 89–105.

4 Erik Swyngedouw and Maria Kaïka, "The Making of 'Glocal' Urban Modernities," *City* 7, no. 1 (April 2003): 10.

5 Many write-ups have appeared in English on Davutoğlu and Turkey's foreign policy. See, for instance, James Traub, "Turkey's Rules," *New York Times*, January 20, 2011, accessed on May 30, 2017, http://www.nytimes.com/.

6 Tanıl Bora, "Istanbul of the Conqueror: The 'Alternative Global City' Dreams of Political Islam," in *Istanbul: Between the Global and the Local*, ed. Çağlar Keyder (Lanham, MD: Rowman & Littlefield, 1999), 48.

7 Cihan Tuğal, "Islam and the Retrenchment of Turkish Conservatism," in *Post-Islamism: The Changing Faces of Political Islam*, ed. Asaf Beyat (Oxford: Oxford University Press, 2013), 109–33.

8 Çiğdem Kafescioğlu, *Constantinopolis/Istanbul, Cultural Encounter, Imperial Vision, and the Construction of the Ottoman Capital* (University Park, PA: Pennsylvania State University Press, 2009).

9 Ibid.

10 I borrow the first term from Eric J. Hobsbawm and Terence O. Ranger's *The Invention of Tradition* (Cambridge; New York: Cambridge University Press, 1983) and the latter from Nezar AlSayyad, ed. *Consuming Tradition, Manufacturing Heritage: Global Norms and Urban Forms in the Age of Tourism* (New York: Routledge, 2001).

11 Çiğdem Kafescioğlu, *Constantinopolis/Istanbul.*

12 For an account that follows print culture, see: Gavin D. Brockett, "Sultan Mehmed II and the Conquest of Constantinople," in *How Happy to Call Oneself a Turk: Provincial Newspapers and the Negotiation of a Muslim National Identity* (Austin: University of Texas Press, 2011), 195–200. For the historical connection of this imagination to contemporary politics, see: Tanıl Bora, "Istanbul of the Conqueror: The 'Alternative Global City' Dreams of Political Islam," in *Istanbul Between the Global and the Local*, ed. Çağlar Keyder (Lanham, MD: Rowman & Littlefield, 1999), 47–58.

13 Martin Stokes, *The Republic of Love: Cultural Intimacy in Turkish Popular Music* (Chicago: University of Chicago Press, 2010), 161–62.

14 Nur Altınyıldız, "The Architectural Heritage of Istanbul and the Ideology of Preservation," *Muqarnas* 23 (2007): 292–3.

15 Istanbul Fetih Cemiyeti's official site: http://www.istanbulfetihcemiyeti.org.tr/
16 Aydın Arakon, dir., *İstanbul'un Fethi 1453* (1951). accessed May 27, 2017, https://youtu.be/uFXMP7u6EM8.
17 Other war films and TV series followed. The most famous of all Ottoman-themed soap operas was *The Magnificent Century*, set during the reign of Sultan Suleiman the Magnificent; it was a broadcast hit not only in Turkey but also in the more than thirty countries in which it was shown.
18 Some of the accounts that connect the panorama to dizzying real estate developments include (but are not limited to): Landon Thomas Jr., "Alarm over Istanbul's Building Boom," *New York Times*, May 20, 2014, accessed May 27, 2017, https://nyti.ms/2snSTDY; Andrew Finkel, "A City Under Siege," *New York Times*, March 1, 2012, accessed May 27, 2017, https://nyti.ms/2rpS2Wm; Dan Bilefsky, "As If the Ottoman Period Never Ended," *New York Times*, Oct 29, 2012, accessed May 27, 2017, https://nyti.ms/2k4Id9s; and Deniz Ünsal, "The Reconquest of Constantinople: Reflections on the Contemporary Landscape and the 1453 Panorama Museum in Istanbul," in *Whose City Is That? Culture, Design, Spectacle and Capital in Istanbul*, eds. Dilek Özhan Koçak and Orhan Kemal Koçak (Newcastle upon Tyne, UK: Cambridge Scholars Publishing, 2014), 283–98.
19 Roland Barthes, *The Eiffel Tower and Other Mythologies*, trans. Richard Howard (New York: Hill and Wang, 1979); and Walter Benjamin, *Arcades Project*, ed. Roolf Tiedemann, trans. Howard Eiland and Kevin McLauglin (Cambridge, MA: Belknap Press of Harvard University Press, 1999), 840.
20 Jonathan Crary, "Géricault, the Panorama, and Sites of Reality in the Early Nineteenth Century," *Grey Room*, no. 9 (Fall 2002): 11.
21 Alison Griffiths, *Shivers Down Your Spine: Cinema, Museums, and the Immersive View* (New York: Columbia University Press, 2008), 41.
22 Brian Massumi, "Autonomy of Affect," *Cultural Critique*, no. 31, The Politics of Systems and Environments, Part II (Autumn, 1995): 84.
23 For Massumi, affect is primary, nonconscious, presubjective, asignifying, unqualified, and intensive. Emotion is derivative, conscious, qualified, and meaningful—a "content" that can be attributed to an already-constituted subject.
24 Benjamin, *Arcades Project*, 840.
25 Bernard Comment, "The Individual in the Town: Compensation and Control," in *The Painted Panorama*, trans. Anne-Marie Glasheen (London: Reaktion Books, 1999), 134–138. Also see Crary, "Géricault, the Panorama, and Sites of Reality," 21.
26 Author's emphasis. Comment, 135.
27 Ayşe Öncü, "The Politics of Istanbul's Ottoman Heritage in the Era of Globalism: Refractions through the Prism of a Theme Park," in *Cities of the South: Citizenship and Exclusion in the 21st Century*, eds. Barbara Drieskens, Franck Mermier, and Heiko Wimmen (London, Beirut: Saqi Books, 2007), 256. Öncü's emphasis.
28 My emphasis. Barthes, *The Eiffel Tower*, 5.
29 Ibid., 11.
30 W. J. T. Mitchell, "Imperial Landscape," in *Landscape and Power* (Chicago: University of Chicago Press, 2002), 10; and Stephan Oettermann, *The Panorama: History of a Mass Medium*, trans. Deborah Lucas Schneider (New York: Zone Books, 1997), 41.
31 Ekrem Işın, "Istanbul Panoramas: Construct, Silhouette, Image," in *Long Stories: Istanbul in the Panoramas of Melling and Dunn*, ed. Ekrem Işın (Istanbul: Istanbul Research Institute, Pera Museum, 2008), 10–37.
32 For a close reading of the shift in representations of the early Ottoman city in relation to the design of the city, see: Kafescioğlu, "Representing the City: Constantinople and Its Images," in *Constantinopolis/Istanbul*, 143–77; 162.
33 Ibid. For a detailed analysis of Lorich's panorama, see: Nigel Westbrook, Kenneth Rainsbury Dark, and Rene Van Meeuwen, "Constructing Melchior Lorichs's Panorama

of Constantinople," *Journal of the Society of Architectural Historians* 69, no. 1 (March 2010): 62–87.
34 Denis Blake Oleksijczuk, *The First Panoramas: Visions of British Imperialism* (Minneapolis, MN: University of Minnesota Press, 2011).
35 Michael Meeker, "Once There Was, Once There Wasn't: National Monuments and Interpersonal Exchange," in *Rethinking Modernity and National Identity in Turkey*, eds. Sibel Bozdoğan and Reşat Kasaba (Seattle, WA: University of Washington Press, 1997), 157–91.
36 Shirine Hamadeh, *The City's Pleasures: Istanbul in the Eighteenth Century* (Seattle, WA: University of Washington Press, 2008).
37 This word is cited in the title of Ralph Hyde, *Panoramania!: The Art and Entertainment of the "All-Embracing"* (London: Trefoil, in association with the Barbican Art Gallery, 1988).
38 Engin Işın, "The Soul of a City: *Hüzün, Keyif*, Longing," in *Orienting Istanbul: Cultural Capital of Europe?*, eds. Deniz Göktürk, Levent Soysal, and Ipek Türeli (London: Routledge, 2010), 35–47.
39 Timothy Mitchell, "The World as Exhibition," *Comparative Studies in Society and History* 31, no. 2 (April 1989), 217–36; 231.
40 Michel de Certeau, "Walking in the City," *The Practice of Everyday Life*, trans. Steven Rendall (Berkeley, CA: University of California Press, 1984), 91–110.
41 Öncü, "The Politics of Istanbul's Ottoman Heritage."
42 See a similar argument about vertical segregation for Athens in: Thomas Maloutas and Nikos Karadimitriou, "Vertical Social Differentiation in Athens: Alternative or Complement to Community Segregation?" *International Journal of Urban and Regional Research* 25, no. 4 (2001): 699–716.
43 Some of companies located there are the confectionary producer Ülker, the home-appliances producer Arçelik, and the Man Truck and Bus. There are also planned industrial zones (*sanayi bölgesi*) the government has urged small manufacturers to move to over the years.
44 Istanbul Metropolitan Municipality website, "İstanbullulara nefes aldıracak 'Topkapı Şehir Parkı' bitmek üzere," last modified July 18, 2006, accessed January 20, 2017, http://www.ibb.gov.tr/tr-TR/Lists/Haber/DispForm.aspx?ID=13224.
45 Istanbul Metropolitan Municipality website, "Belediye Meclis Kararları: 937," accessed July 17, 2009, http://www.ibb.gov.tr/tr-TR/Lists/BelediyeMeclisKararlari/DispForm.aspx?ID=15645.
46 For a study of the park, see: Helin Karaman, "Le Topkapı Kültür Parkı: fabriquer un parc public à Istanbul," *European Journal of Turkish Studies*, no. 23 (2016), accessed January 2, 2017. http://ejts.revues.org/5389.
47 My emphasis. Zeytinburnu municipal website: "1998 yılında eski Trakya Otogarı yerinde Topkapı Şehir Parkı ile başlayan projeler dizisi ise, yörenin çehresini orta ve uzun vadede değiştirebilme potansiyelini barındırmaktadır. Daha ileri aşamalarda Surlar boyunca uzanan ve dini turizmi destekleyen bir kültür adası yaratılması hedeflenmektedir." Accessed July 17, 2009, http://www.zeytinburnu.com.tr/Sayfa/83/tarihce/modern-zeytinburnu.aspx.
48 Zinnet Restaurant's website explains that its *raison d'être* was "anthropological": "Bir ulusun yemek kültürü ise, o ulusun antropolojik özelliklerini en iyi şekilde ifade eden önemli unsurdur" (A nation's culinary culture is the best expression of that nation's anthropological characteristics). Accessed July 17, 2009, http://www.zinnetrestaurant.com/hakkimizda.htm.
49 "Topkapı'da Osmanlı Köyü Kuruluyor," *Arkitera.com*, March 5, 2003, accessed May 27, 2017, http://v3.arkitera.com/v1/haberler/2003/03/05/topkapi.htm.
50 Newroz (or Nevruz) celebrations in Turkey are associated with Kurdish nationalism, and its celebration has been suppressed and then co-opted by the Turkish state since the 1990s. However, it is a special day celebrated in late March across the region from the

Balkans to Central Asia. It is the first day of the year and the start of the spring in the Persian tradition. For Alawite-Bektashi communities, it is also celebrated as the birthday of Ali. Lerna K. Yanik, "'Nevruz' or 'Newroz'? Deconstructing the 'Invention' of a Contested Tradition in Contemporary Turkey," *Journal of Middle Eastern Studies* 42, no. 2 (March 2006): 285–302.

51 Ministry of Foreign Affairs of the Republic of Turkey, official website, "Turkey's Relations with Central Asian Republics," accessed July 17, 2009, http://www.mfa.gov.tr/turkey_s-relations-with-central-asian-republics.en.mfa.

52 These figures were provided to me by the operational chief (manager) during an interview on August 8, 2011.

53 This was İbrahim Karagül ve Hüsnü Mahalli's "Büyük Oyun" that appeared on TV Net and was later moved to the studio setting because the panorama setting apparently distracted both the producers and the viewers of the program.

54 Interviews conducted on August 8, 2011. While I was not able to identify an architectural firm, two construction firms, Bilal İnşaat and Doğan İnşaat, have listed the project on their websites. Accessed August 10, 2011, http://www.biatinsaat.com.tr/ and http://www.doganinsaattaahhut.com.

55 Carol Duncan, "The Art Museum as Ritual," *Civilizing Rituals: Inside Public Art Museums* (London; New York: Routledge, 1995), 7–20.

56 Griffiths, *Shivers Down Your Spine*, 41.

57 For example: Korhan Gümüş, "Fetih Müzesi Neyi Simgeliyor?" *Radikal*, June 3, 2009, accessed January 10, 2017. http://v3.arkitera.com/h41818-fetih-muzesi-neyi-simgeliyor.html.

58 Thomas Campanella, *The Concrete Dragon: China's Urban Revolution and What It Means for the World* (New York: Princeton Architectural Press, 2008).

59 Some of the pioneering studies of panoramas include Richard D. Altick, *The Shows of London* (Cambridge, MA: The Belknap Press of Harvard University Press, 1978); Oettermann, *The Panorama*; and Hyde, *Panoramania!*

60 The official website of the organization is http://panoramacouncil.org/. *Panorama 1453/Panorama Museum* DVD issued by Istanbul Metropolitan Municipality and its subsidiary Kültür A.Ş.

61 For example, Lord Nelson's Defeat of the French on the Nile, 1798. A contemporary example is the October Panorama (Tishreen Al-'Awal), a Damascus monument to the war beginning October 1973, also referred to as the Yom Kippur War.

62 Korean artists have carved a niche for themselves internationally: The panoramas of Damascus and Cairo were painted by them.

63 Ottermann, *The Panorama*, 113.

64 Dietrich Neumann, "Instead of the Grand Tour: Travel Replacements in the Nineteenth Century," *Perspecta* 5, no. 41 (2009): 47–53; and Andreas Luescher, "Great Travel Machines of Sight," in *Travel, Space, Architecture*, eds. Jilly Traganou and Miodrag Mitrašinović (Farnham, UK: Ashgate, 2009), 48–63.

65 The first panorama was painted by Robert Barker's son Henry Aston Barker. Reinhold Shiffer, "Panoramic Views of Istanbul," *Oriental Panorama: British Travellers in 19th Century Turkey* (Atlanta, GA: Rodopi, 1999), 145–8; Denis Blake Oleksijczuk, "Chapter 4: The Views of Constantinople," *The First Panoramas: Visions of British Imperialism* (Minneapolis, MN: University of Minnesota Press, 2011), 89–125; Namık Erkal, "Tam Zamanında Gözlerinizin Önünde: Londra Panoramalarında İstanbul Sergileri" (part 1), *Toplumsal Tarih*, no. 170 (2008): 40–7; (part 2) no. 171 (2008): 24–31.

66 Shiffer, "Panoramic Views of Istanbul," 145–6.

67 Kültür A.Ş.'s (former) Chairman Nevzat Baykan in the promotional DVD for the panorama. Also reported by the panorama's artistic coordinator in my interview.

68 Griffiths, *Shivers Down Your Spine*, 41.

69 İlber Ortaylı is featured in the promotional DVD for the panorama.

70 Sky Ride Istanbul. "Sky Ride Istanbul," accessed January 20, 2017. www.skyride.com.tr.

Bibliography

AlSayyad, Nezar, ed. *Consuming Tradition, Manufacturing Heritage: Global Norms and Urban Forms in the Age of Tourism*. New York: Routledge, 2001.

Altick, Richard D. *The Shows of London*. Cambridge, MA: The Belknap Press of Harvard University Press, 1978.

Altınyıldız, Nur. "The Architectural Heritage of Istanbul and the Ideology of Preservation." *Muqarnas* 23 (2007): 292–3.

Barthes, Roland. *The Eiffel Tower and Other Mythologies*, translated by Richard Howard. New York: Hill and Wang, 1979.

Benjamin, Walter. *Arcades Project*, edited by Roolf Tiedemann, translated by Howard Eiland and Kevin McLauglin. Cambridge, MA: Belknap Press of Harvard University Press, 1999.

Bennett, Tony. *The Birth of the Museum: History, Theory, Politics*. London; New York: Routledge, 1995.

Bilefsky, Dan. "As If the Ottoman Period Never Ended." *New York Times*, Oct 29, 2012. Accessed May 27, 2017. https://nyti.ms/2k4Id9s.

Bora, Tanıl. "Istanbul of the Conqueror: The 'Alternative Global City' Dreams of Political Islam." In *Istanbul: Between the Global and the Local*, edited by Çağlar Keyder, 47–58. Lanham, MD: Rowman & Littlefield, 1999.

Brockett, Gavin D. *How Happy to Call Oneself a Turk: Provincial Newspapers and the Negotiation of a Muslim National Identity*. Austin: University of Texas Press, 2011.

Campanella, Thomas. *The Concrete Dragon: China's Urban Revolution and What It Means for the World*. New York: Princeton Architectural Press, 2008.

de Certeau, Michel. *The Practice of Everyday Life*, translated by Steven Rendall. Berkeley, CA: University of California Press, 1984.

Comment, Bernard. *The Painted Panorama*, translated by Anne-Marie Glasheen. London: Reaktion Books, 1999.

Crary, Jonathan. "Géricault, the Panorama, and Sites of Reality in the Early Nineteenth Century." *Grey Room*, no. 9 (Fall 2002): 5–25.

Duncan, Carol. *Civilizing Rituals: Inside Public Art Museums*. London; New York: Routledge, 1995.

Erkal, Namık. "Tam Zamanında Gözlerinizin Önünde: Londra Panoramalarında İstanbul Sergileri"; (part 1) *Toplumsal Tarih*, no. 170 (2008): 40–7; (part 2) no. 171 (2008): 24–31.

Finkel, Andrew. "A City Under Siege." *New York Times*, March 1, 2012. Accessed May 27, 2017. https://nyti.ms/2rpS2Wm.

Griffiths, Alison. *Shivers Down Your Spine: Cinema, Museums, and the Immersive View*. New York: Columbia University Press, 2008.

Gül, Abdullah. 15 March 2009, Guest Book. Reproduced in *Panorama 1453 Tarih Müzesi, Panorama 1453 Historical Museum*. Istanbul: İstanbul Büyük Şehir Belediyesi, 2009.

Gümüş, Korhan. "Fetih Müzesi Neyi Simgeliyor?" *Radikal*, June 3, 2009. Accessed January 10, 2017. http://v3.arkitera.com/h41818-fetih-muzesi-neyi-simgeliyor.html.

Hamadeh, Shirine. *The City's Pleasures: Istanbul in the Eighteenth Century*. Seattle, WA: University of Washington Press, 2008.

Hobsbawm, Eric J., and Terence O. Ranger. *The Invention of Tradition*. Cambridge; New York: Cambridge University Press, 1983.

Hyde, Ralph. *Panoramania!: The Art and Entertainment of the "All-Embracing."* London: Trefoil, in association with the Barbican Art Gallery, 1988.

Işın, Ekrem, ed. *Long Stories: Istanbul in the Panoramas of Melling and Dunn*. Istanbul: Istanbul Research Institute, Pera Museum, 2008.

Işın, Engin. "The Soul of a City: *Hüzün, Keyif*, Longing." In *Orienting Istanbul: Cultural Capital of Europe?*, edited by Deniz Göktürk, Levent Soysal, and Ipek Türeli, 35–47. London: Routledge, 2010.

Istanbul Metropolitan Municipality website. "Belediye Meclis Kararları: 937." Accessed July 17, 2009. http://www.ibb.gov.tr/tr-TR/Lists/BelediyeMeclisKararlari/DispForm.aspx?ID=15645

—. "İstanbullulara nefes aldıracak 'Topkapı Şehir Parkı' bitmek üzere." July 18, 2006. Accessed January 20, 2017. http://www.ibb.gov.tr/tr-TR/Lists/Haber/DispForm.aspx?ID=13224.

—. "Türkiye'nin ilk panoramik müzesi 'İstanbul'un Fethi'ni' yeniden yaşatacak." January 31, 2009. Accessed on May 29, 2017, http://www.ibb.gov.tr/tr-TR/Lists/Haber/DispForm.aspx?ID=17004.

Kafescioğlu, Çiğdem. *Constantinopolis/Istanbul, Cultural Encounter, Imperial Vision, and the Construction of the Ottoman Capital*. University Park, PA: Pennsylvania State University Press, 2009.

Karaman, Helin. "Le Topkapı Kültür Parkı: fabriquer un parc public à Istanbul." *European Journal of Turkish Studies*, no. 23 (2016). Accessed on January 2, 2017. http://ejts.revues.org/5389.

Luescher, Andreas. "Great Travel Machines of Sight." In *Travel, Space, Architecture*, edited by Jilly Traganou and Miodrag Mitrašinović, 48–63. Farnham, UK: Ashgate, 2009.

Maloutas, Thomas, and Nikos Karadimitriou, "Vertical Social Differentiation in Athens: Alternative or Complement to Community Segregation?" *International Journal of Urban and Regional Research* 25, no. 4 (2001): 699–716.

Massumi, Brian. "Autonomy of Affect." *Cultural Critique*, no. 31, The Politics of Systems and Environments, Part II (Autumn, 1995): 83–109.

Meeker, Michael. "Once There Was, Once There Wasn't: National Monuments and Interpersonal Exchange." In *Rethinking Modernity and National Identity in Turkey*, eds. Sibel Bozdoğan and Reşat Kasaba, 157–91. Seattle, WA: University of Washington Press, 1997.

Ministry of Foreign Affairs of the Republic of Turkey. "Turkey's Relations with Central Asian Republics." Accessed of July 17, 2009. http://www.mfa.gov.tr/turkey_s-relations-withcentral-asian-republics.en.mfa.

Mitchell, Timothy. "The World as Exhibition." *Comparative Studies in Society and History* 31, no. 2 (April 1989): 217–36.

Mitchell, W. J. T., ed. *Landscape and Power*. Chicago, IL: University of Chicago Press, 2002.

Neumann, Dietrich. "Instead of the Grand Tour: Travel Replacements in the Nineteenth Century." *Perspecta* 5, no. 41 (2009): 47–53.

Oettermann, Stephan. *The Panorama: History of a Mass Medium*, translated by Deborah Lucas Schneider. New York: Zone Books, 1997.

Oleksijczuk, Denis Blake. *The First Panoramas: Visions of British Imperialism*. Minneapolis, MN: University of Minnesota Press, 2011.

Öncü, Ayşe. "The Politics of Istanbul's Ottoman Heritage in the Era of Globalism: Refractions through the Prism of a Theme Park." In *Cities of the South: Citizenship and Exclusion in the 21st Century*, edited by Barbara Drieskens, Franck Mermier, and Heiko Wimmen, 233–64. London; Beirut: Saqi Books, 2007.

Shiffer, Reinhold. *Oriental Panorama: British Travellers in 19th Century Turkey*. Atlanta, GA: Rodopi, 1999.
Sky Ride Istanbul. "Sky Ride Istanbul." Accessed January 20, 2017. www.skyride.com.tr.
Stokes, Martin. *The Republic of Love: Cultural Intimacy in Turkish Popular Music*. Chicago: University of Chicago Press, 2010.
Swyngedouw, Erik, and Maria Kaïka, "The Making of 'Glocal' Urban Modernities." *City* 7, no. 1 (April 2003): 5–21.
Thomas Jr., Landon. "Alarm over Istanbul's Building Boom." *New York Times*, May 20, 2014. Accessed May 27, 2017. https://nyti.ms/2snSTDY.
"Topkapı'da Osmanlı Köyü Kuruluyor." *Arkitera.com*, March 5, 2003, accessed May 27, 2017, http://v3.arkitera.com/v1/haberler/2003/03/05/topkapi.htm.
Traub, James. "Turkey's Rules." *New York Times*, January 20, 2011. Accessed May 27, 2017. https://nyti.ms/2oyxXfy..
Tuğal, Cihan. "Islam and the Retrenchment of Turkish Conservatism." In *Post-Islamism: The Changing Faces of Political Islam*, edited by Asaf Beyat, 109–33. Oxford: Oxford University Press, 2013.
Ünsal, Deniz. "The Reconquest of Constantinople: Reflections on the Contemporary Landscape and the 1453 Panorama Museum in Istanbul." In *Whose City Is That? Culture, Design, Spectacle and Capital in Istanbul*, edited by Dilek Özhan Koçak and Orhan Kemal Koçak, 283–98. Newcastle upon Tyne, UK: Cambridge Scholars Publishing, 2014.
Westbrook, Nigel, Kenneth Rainsbury Dark, and Rene Van Meeuwen. "Constructing Melchior Lorichs's Panorama of Constantinople." *Journal of the Society of Architectural Historians* 69, no. 1 (March 2010): 62–87.
Yanik, Lerna K. "'Nevruz' or 'Newroz'? Deconstructing the 'Invention' of a Contested Tradition in Contemporary Turkey." *Journal of Middle Eastern Studies* 42, no. 2 (March 2006): 285–302.

7 Conclusion

Refuge in the open city

> "the history and culture of each city is sufficiently different to require, in addition to the more basic analytical concepts, its own descriptions and categories drawn from its own cultural histories and languages."[1]

The basic analytical concept at the center of this book is the "open city." Acceptance of population influx, as well as lack of planning controls, has been central to different articulations of the concept. While the development of the urban form of the city since the 1950s has been marked by leniency of planning controls, various governments' episodic heavy-handed urban interventions have tended to exacerbate anxieties about the city. Such urban interventions became a tool of national politics and literally the focus of national election campaigns: during the latter part of the 1950s under the Democrat Party, the latter part of the 1980s under the Motherland Party, and since 2002, under the AKP. In the first two, large swathes of the city were transformed by drastic and piecemeal projects, mainly concerning roads, that promoted motor vehicles and helped the city geographically spread with new housing projects. During the AKP years, urban transformation has increased on an enormous scale due to the availability of new types of international lending and public private partnerships, as well as relative stability due to the continuity of the AKP's rule, which is now in its fifteenth year—the longest since the single-party era of the early Republic. All of these episodes were characterized by the opening of the city, in reference to the physical opening of boulevards through the historic fabric, demographic changes, and an inflow of foreign goods and ideas due to market reforms.

While "open city" is a contemporary concept, I have pointed out how the dramatic acceleration of internal migration to Istanbul in the 1950s led to rapid population growth, facilitating the city's physical expansion, and served as a justification for urban renewal, which in fact led contemporary commentators in the press to suggest that the city had become an open city. I have suggested that this double movement, population influx and physical outward expansion, led to an obsession with the city in the realm of culture that has persisted to the present day. I have argued that the media representations of the city I discussed have served to alleviate anxieties related to the city's growth and transformation.

I have noted that these representations were often portrayed as escape or refuge from the contemporary city—from the complexity and contradictions of everyday experience in the city or from problems associated with it. Yet—and this is the distinction of my argument—in the way they edit and collage history, media representations of the city improvise on the past; they act as prosthetic memories that allow both their producers and viewers new, flexible ways of defining themselves; and most importantly, they enable public debate about the future of the city.

Destination-city and the migrant

The rural-to-urban migrant emerged as an important figure of urban modernity early on, as depicted in cartoons, photographs for print journalism, and popular cinema films. Chapters 2 and 3, on print press/photography and cinema/films, dealt more explicitly with the figure of the migrant, in whose mirror image the Istanbulite was construed. Journalists, and filmmakers inspired by journalists' reports, acknowledged middle-class fears and concerns over the limits of the city's resources, but they also reasoned that the influx to Istanbul was not new in the city's long history. It was necessary, in fact, for the city's projected industrialization, as the migrants provided a much needed labor pool. Nevertheless, economically driven internal migration had a profound effect on the culture of the city and gave way to a verbose discourse about "integration" (or the failure thereof) within Turkey and to anxieties about the "provincialization" of Istanbul. Although no longer the driving force of the city's population growth, migration remains central to cultural imagination. This cultural imagination has an underlying assumption: that Istanbul is a destination city.

What this book has not touched upon is that from the mid-1980s, the civil war in southeast Turkey led to the forced migration of Kurds into the city, complicating the general perception of migration and the figure of the migrant. In addition, since the 1990s, international migration and mobility have brought in waves from post-Soviet countries, conflict zones in the broader region, and African countries. Increasing labor flexibility in the local market has led to employers preferring lower-paid international migrants. Some of these are economic migrants who come to Turkey for short periods to send remittances back home. Many are individuals seeking refuge—but not necessarily in Istanbul. In the documentary *Callshop Istanbul* (2016, dir. Hind Benchekroun and Sami Mermer of Montreal, Quebec), Istanbul is portrayed through the long-distance phone calls made by refugees, immigrants, and drifters in call shops. Coming from a diverse range of backgrounds and geographies—including, but not limited to, Benin, Iraq, Iran, Japan, Senegal, Spain, and Syria—what unites these individuals is that they are passing through Istanbul, and none of them seems to want to make it their own or call it their home. Istanbul has emerged in global networks but has denied being part of Europe. Thus, in a time of a global forced migration crisis, the city has transformed, especially for international refugees, into a "waiting zone," rather than a destination.

Museum-city and the city enthusiast

The exhibitions of old Istanbul houses and streets that were discussed in Chapter 4 evidence the emergence of a new figure, the Istanbul enthusiast, who seeks refuge from rampant urbanization, marked by the proliferation of speculative concrete-frame apartment buildings and the disappearance of historic wooden ones. In doing so, they make their own path while simultaneously following in the footsteps of an earlier generation which dwelled on "Turkish Istanbul" as a critique of the Republican project of modernization. Istanbul enthusiasts exhibited old houses, streetscapes, and neighborhoods without any reference to their departed former residents and with overt nationalistic justifications. I have also highlighted that these exhibitions and calls to conserve old houses were in line with international developments and institutional frameworks for tourism-oriented area conservation and, thus, readily supported by the Turkish government. The romanticization of old houses in the discussed exhibitions lacked historicization, and as such lent themselves easily to media productions in the following decades, including cinema films and numerous popular TV serials. These screen-based productions effectively used the spatial characteristics of old neighborhoods as outdoor stages and further promoted the gentrification of the old neighborhoods in which they were shot. To emphasize, the revalorization of old houses and calls for their conservation was coeval and in conversation with imaginings of the commodified museum-city in global metropoles since the 1970s as a booster-strategy to revive urban economies, first via tourism and later via creative sectors.

As the historic city was turned into an open museum (beginning with installations such as Soğukçeşme Street), the notion of seeking refuge from the city of apartment blocks became mainstream. It became a trope in the advertisement of the city's new housing developments to contrast the gray (concrete), polluted, and supposedly dangerous city with green and socially homogeneous suburban developments. Both the proliferation of exclusive gated communities and more affordable mass housing *sites* have received due attention in the sociological scholarship on Istanbul. Class-based housing segregation as refuge from the city is antithetical to the ideal of the open city but not necessarily novel; it is indeed one of the myriad ways the city seeks to control and manage increasing cultural complexity.

From the 1990s, the government took a new interest in turning Istanbul into a cultural capital, in collaboration with a new host of civil society organizations and philanthropic institutions invested in a new generation of private museums. Istanbul came to be seen as "Turkey's passport into Europe." Effectively, then, the work of the enthusiast was co-opted. The government promoted Istanbul as a brand while using multiculturalism as a strategy for the management of difference. In its journals, on its billboards around the city, and even with a distinct campaign, the municipality advocated "Istanbulite consciousness" as an identity that could supplant other types of regional attachment. On the one hand, prominent civil society organizations—e.g. the History Foundation—similarly promoted Istanbuliteness as city attachment and initiated efforts to write the history of the city as a multicultural (multi-religious) one—despite the difficulty of addressing the forced departures of the city's non-Muslim communities. On the other hand, a new generation of associations—e.g. Arnavutköy District Initiative against the third bridge construction in 1998—sprang up in the face of specific, heavy-handed urban transformation projects; these organized not only around "culture" but also "environment."

Simulation-city

The introduction briefly explained that this book would explore different "phases of the image," beginning with the reflection of the urban condition in photojournalism, continuing with its fictionalization in film, then its nostalgic return in Old Istanbul exhibitions, and, finally, its simulacrum in theme park and museum settings. The "phases of the image," a concept borrowed from Jean Baudrillard, was particularly useful in conceptualizing the different exhibitionary sites examined in this book and the modalities in which they related to urban change. Press photography, for instance, was always highly framed and edited by the photographer and the editors. In reaching out to its viewers, the individual photograph claimed faithful reproduction, but viewers inevitably read alternative narratives from the same photograph. Over time, these readings changed dramatically. Thus, as I have discussed, while the photograph claimed authenticity, this authenticity was always undermined by the different ways it was coded and the different contexts in which it circulated—from the pages of newspapers to art galleries to coffee-table books to café decoration. Films, on the other hand, were staged narratives that had the effect of fictionalizing and spectacularizing everyday lived experience. As I have discussed, the process of fictionalization offered the possibility of challenging and altering existing power hierarchies. Old Istanbul exhibitions sought to retrieve an imagined past as a means to challenge the status of the present.

What is unique of post-2002, the AKP-era urban imagination is how globally circulating formulas of neoliberal urban policies and mega-projects have been adapted to the local context with a modernist belief in the uplifting, educational effect of culture. Thus, I allocated Chapters 5 and 6 to discuss cultural attractions produced at two different points during the ruling party's evolving political orientation.

The miniature park promised to edit history and bring it to life. It did not show the historical lived experience (of a multicultural past), but rather made it "present" by flattening that history, decontextualizing the models and presenting them in arbitrary arrangements. Seeing the city in miniature, and able to walk around the dispersed architectural models, the viewers of the miniature theme park assume the point of view of the modernist city planner, as if they could be participants in the city's reconfiguration. Yet, the city, as a background to the models, reminded them of the artifice of the display.

Chapter 6, on Panorama 1453, shows how the shifting idea and ideal of conquest serves the needs of the present, as Istanbul is being reconquered to be marketed to global capital. The immersive view demands visitors to assume the point of view of the conquering soldiers on the ground. Yet, they were able to augment this "'old" media with their own technology, zooming in and out using their cameras, thereby extending the given view; the platform's space, though restricted, doubled as a platform for socialization and conversation on the version of urban history displayed.

The repeated emphasis of the relevance, recycling, and recirculation of earlier representations of the city in the present negates the possibility of thinking of a linear development from one representational mode to the other. Although I suggested a trajectory, it involved a shift from progress-oriented to nostalgia-oriented urban modernity, in response to very specific local conditions that were informed and shaped by state-led efforts and global flows, but not necessarily determined by them. The objective of examining different exhibitionary sites was to better understand the effects and implications of aesthetic and spatial strategies by way of comparison.

The exhibitionary sites discussed were chosen because of their capacity to highlight the ephemeral reflections, concerns, and debates experienced by local publics in response to the changes and demands of life in a fast urbanizing city. These sites were strategically crafted—through the framing and captioning of photographs, the editing and framing of films, or the curation of architectural artifacts and models—to communicate narratives of the past in order to imagine the future of the city. They presented their producers and viewers with opportunities to frame the future of their city in new ways, and in doing so, they also demand them to take positions.

Note

1 Anthony D. King, "Boundaries, Networks, and Cities: Playing and Replaying Diasporas and Histories," in *Urban Imaginaries: Locating the Modern City*, eds. Alev Çınar and Thomas Bender (Minneapolis, MN: University of Minnesota Press, 2007), 11–12.

Bibliography

King, Anthony D. "Boundaries, Networks, and Cities: Playing and Replaying Diasporas and Histories." In *Urban Imaginaries: Locating the Modern City*, edited by Alev Çınar and Thomas Bender, 1–14. Minneapolis, MN: University of Minnesota Press, 2007.

Index

For Product Safety Concerns and Information please contact our EU representative GPSR@taylorandfrancis.com Taylor & Francis Verlag GmbH, Kaufingerstraße 24, 80331 München, Germany

Batch number: 10725973

Printed by Printforce, the Netherlands